A Gyrovector Space
Approach to
Hyperbolic Geometry

SYNTHESIS LECTURES ON MATHEMATICS AND STATISTICS

Editor
Steven G. Krantz, *Washington University, St. Louis*

A Gyrovector Space Approach to Hyperbolic Geometry
Abraham Albert Ungar

ISBN: 978-3-031-01268-6   paperback

ISBN: 978-3-031-02396-5   ebook

DOI 10.1007/978-3-031-02396-5

A Publication in the Springer series
SYNTHESIS LECTURES ON MATHEMATICS AND STATISTICS

Lecture #4
Series Editor: Steven G. Krantz, *Washington University, St. Louis*

Series ISSN
Synthesis Lectures on Mathematics and Statistics   Print 1930-1743   Electronic 1930-1751

A Gyrovector Space
Approach to
Hyperbolic Geometry

Abraham Albert Ungar
North Dakota State University

SYNTHESIS LECTURES ON MATHEMATICS AND STATISTICS #4

ABSTRACT

The mere mention of hyperbolic geometry is enough to strike fear in the heart of the undergraduate mathematics and physics student. Some regard themselves as excluded from the profound insights of hyperbolic geometry so that this enormous portion of human achievement is a closed door to them. The mission of this book is to open that door by making the hyperbolic geometry of Bolyai and Lobachevsky, as well as the special relativity theory of Einstein that it regulates, accessible to a wider audience in terms of novel analogies that the modern and unknown share with the classical and familiar.

These novel analogies that this book captures stem from Thomas gyration, which is the mathematical abstraction of the relativistic effect known as Thomas precession. Remarkably, the mere introduction of Thomas gyration turns Euclidean geometry into hyperbolic geometry, and reveals mystique analogies that the two geometries share. Accordingly, Thomas gyration gives rise to the prefix "gyro" that is extensively used in the gyrolanguage of this book, giving rise to terms like gyrocommutative and gyroassociative binary operations in gyrogroups, and gyrovectors in gyrovector spaces. Of particular importance is the introduction of gyrovectors into hyperbolic geometry, where they are equivalence classes that add according to the gyroparallelogram law in full analogy with vectors, which are equivalence classes that add according to the parallelogram law. A gyroparallelogram, in turn, is a gyroquadrilateral the two gyrodiagonals of which intersect at their gyromidpoints in full analogy with a parallelogram, which is a quadrilateral the two diagonals of which intersect at their midpoints.

KEYWORDS

hyperbolic geometry, Analogies between Euclidean and hyperbolic geometry, Poincaré ball model of hyperbolic geometry, Beltrami-Klein ball model of hyperbolic geometry, Möbius transformations of the complex open unit disc, Möbius addition, from Möbius to gyrogroups, gyrogroups, gyrocommutative gyrogroups, gyrovectors, gyrovector spaces, gyrotrigonometry, Einstein relativistic velocity addition, relativistic stellar aberration, dark matter

Contents

Preface

This book is an outgrowth of the success of the theory of gyrogroups and gyrovector spaces in revealing novel analogies in its forerunners [54, 55, 61]. It seeks to provide a gyrovector space approach to hyperbolic geometry, which is fully analogous to the common vector space approach to Euclidean geometry.

Hyperbolic geometry didn't come about just to make things difficult. Rather, it arose because the universe itself, according to Einstein's special theory of relativity, is essentially regulated by hyperbolic geometry. The mere mention of hyperbolic geometry is enough to strike fear in the heart of the undergraduate mathematics and physics student. Some regard themselves as excluded from the profound insights of hyperbolic geometry so that this enormous portion of human achievement is a closed door to them. The mission of this book is to open that door by making the hyperbolic geometry of Bolyai and Lobachevsky, as well as the special relativity theory of Einstein that it regulates, accessible to a wider audience in terms of novel analogies that the modern and unknown share with the classical and familiar.

Non-Euclidean geometry, the hyperbolic geometry of Bolyai and Lobachevsky, resulted from the denial of Euclid's fifth postulate, the parallel postulate, and as such, it possesses far more degrees of freedom than Euclidean geometry. It was created in the first half of the nineteenth century in the midst of attempts to understand Euclid's axiomatic basis for geometry [29]. In the early part of the twentieth century, every serious student of mathematics and physics studied hyperbolic geometry. This has not been true of the mathematicians and physicists of our generation, as Cannon, Floyd, Kenyon and Parry testify in [4]. The decline of interest in the hyperbolic geometry of Bolyai and Lobachevsky began in 1912, as the historian of relativity physics, Scott Walter, explains:

> Over the years, there have been a handful of attempts to promote the non-Euclidean style for use in problem solving in relativity and electrodynamics, the failure of which to attract any substantial following, compounded by the absence of any positive results must give pause to anyone considering a similar undertaking. Until recently, no one was in a position to offer an improvement on the tools available since 1912. In his [2001] book, Ungar furnishes the crucial missing element from the panoply of the non-Euclidean style: an elegant nonassociative algebraic formalism that fully exploits the structure of Einstein's law of velocity composition. The formalism relies on what the author calls the "missing link" between Einstein's velocity addition formula and ordinary vector addition: Thomas precession …

Scott Walter
Foundations of Physics 32, pp. 327 – 330 (2002)

Indeed, *Thomas precession* is the missing link between hyperbolic and Euclidean geometry. It is a relativistic effect whose mathematical abstraction is called *Thomas gyration*. The latter, in turn, gives rise to the prefix "gyro" that we extensively use in the *gyrolanguage* of this book, leading to terms like gyrocommutative and gyroassociative binary operations in gyrogroups, and gyrovectors in gyrovector spaces. Remarkably, the mere introduction of Thomas gyration turns Euclidean geometry into hyperbolic geometry, and reveals mystique analogies that the two geometries share.

Accordingly, in the four chapters of this book we present a powerful way to study the hyperbolic geometry of Bolyai and Lobachevsky, in which analogies with Euclidean geometry form the right tool. In this book, we offer the radical break of the traditional teaching of hyperbolic geometry required to restore its glory and harmony, ascending it to levels that place it in the mainstream of mathematics and physics. We thus demonstrate in this book that the time of the hyperbolic geometry of Bolyai and Lobachevsky to be part of the lore of every serious student of mathematics and physics has returned.

In Chapter 1, entitled *"Gyrogroups"*, we motivate the generalization of the notion of the group into that of the gyrogroup, and show that the mere introduction of the notion of the *gyration* turns groups into gyrogroups. In the same way that some groups are commutative, some gyrogroups are gyrocommutative.

In Chapter 2, entitled *"Gyrocommutative Gyrogroups"*, we study gyrocommutative gyrogroups. Their importance stems from the result that some gyrocommutative gyrogroups admit scalar multiplication that gives rise to gyrovector spaces just as some commutative groups admit scalar multiplication that gives rise to vector spaces. Furthermore, the abstract gyrocommutative gyrogroup is realized in this chapter by two concrete examples, the Möbius gyrogroups in Section 2.2 and the Einstein gyrogroups in Section 2.3.

In Chapter 3, entitled *"Gyrovector Spaces"*, we present and study gyrovector spaces. In particular, we add structure to Möbius gyrogroups and Einstein gyrogroups by incorporating scalar multiplication, resulting in Möbius gyrovector spaces and Einstein gyrovector spaces. Remarkably, gyrovector spaces form the algebraic setting for the hyperbolic geometry of Bolyai and Lobachevsky just as vector spaces form the algebraic setting for the standard model of Euclidean geometry. In particular, Möbius gyrovector spaces form the algebraic setting for the Poincaré ball model of hyperbolic geometry and, similarly, Einstein gyrovector spaces form the algebraic setting for the Beltrami-Klein ball model of hyperbolic geometry. In this chapter, accordingly, our gyrovector space approach to hyperbolic geometry emerges in full analogy with the common vector space approach to Euclidean geometry. Thus, for instance, gyrovectors turn out to be equivalence classes that add according to the gyroparallelogram law in full analogy with vectors, which are equivalence classes that add according to the parallelogram law. A gyroparallelogram, in turn, is a gyroquadrilateral the two gyrodiagonals of which intersect at their gyromidpoints in full analogy with a parallelogram, which is a quadrilateral the two diagonals of which intersect at their midpoints.

In Chapter 4, entitled *"Gyrotrigonometry"*, we present and study the gyrotrigonometry of gyrovector spaces in full analogy with the common trigonometry of vector spaces. In fact, the analogies work so well that the basic trigonometric functions $\sin\alpha$, $\cos\alpha$, etc., share their functional relationships, like $\sin^2\alpha + \cos^2\alpha = 1$, with the basic gyrotrigonometric functions, also denoted by $\sin\alpha$, $\cos\alpha$, *etc.* Most importantly, we demonstrate in this chapter how results in Euclidean geometry are translated by means of our gyrogeometric techniques into novel results in hyperbolic geometry. Readers of this book are therefore likely to be able to prove their own theorems in hyperbolic geometry by translating some known theorems in Euclidean geometry.

This book, thus, culminates in a most inspirational vision. The hyperbolic geometry of Bolyai and Lobachevsky, as studied in this book, offers great promise for more discoveries in the gyrovector space approach to hyperbolic geometry to come.

As a mathematical prerequisite for a fruitful reading of this book it is assumed familiarity with the common vector space approach to Euclidean geometry.

Abraham Albert Ungar
North Dakota State University
Fargo, ND USA
January 2009

CHAPTER 1

Gyrogroups

Gyrogroups are generalized groups, which are best motivated by the algebra of Möbius transformations of the complex open unit disc. Groups are classified into commutative and non-commutative groups and, in full analogy, gyrogroups are classified into gyrocommutative and non-gyrocommutative gyrogroups. Some commutative groups admit scalar multiplication, giving rise to vector spaces. In full analogy, some gyrocommutative gyrogroups admit scalar multiplication, giving rise to gyrovector spaces. Furthermore, vector spaces form the algebraic setting for the standard model of Euclidean geometry and, in full analogy, gyrovector spaces form the algebraic setting for various models of the hyperbolic geometry of Bolyai and Lobachevsky.

The evolution from Möbius to gyrogroups began soon after the discovery in 1988 [52] that Einstein velocity addition law of relativistically admissible velocities, which is seemingly structureless, is in fact rich of structure. Later, the rich structure that Einstein velocity addition law encodes turned out to be a gyrocommutative gyrogroup and a gyrovector space. The resulting notions of gyrogroups and gyrovector spaces preserve the flavor of their classical counterparts, groups and vector spaces. They are useful and fascinating enough to be made part of the lore learned by all undergraduate and graduate mathematics and physics students. Being a natural generalization of groups and vector spaces, gyrogroups and gyrovector spaces lay a fruitful bridge between nonassociative algebra and hyperbolic geometry, just as groups and vector spaces lay a fruitful bridge between associative algebra and Euclidean geometry. More than 150 years have passed since the German mathematician August Ferdinand Möbius (1790–1868) first studied the transformations that now bear his name [33, p. 71]. Yet, the rich structure he thereby exposed is still far from being exhausted, as the evolution from Möbius to gyrogroups demonstrates in Sections 1.1–1.3 of this chapter.

1.1 FROM MÖBIUS TO GYROGROUPS

Möbius transformations of the complex open unit disc $\mathbb{D}$,

$$\mathbb{D} = \{z \in \mathbb{C} : |z| < 1\} \tag{1.1}$$

of the complex plane $\mathbb{C}$ are studied in most books on function theory of one complex variable. According to [35, p. 176], these are important for at least two reasons: (i) they play a central role in non-Euclidean geometry [55], and (ii) they are the only automorphisms of the disc.

Ahlfors' book [1], *Conformal Invariants: Topics in Geometric Function Theory*, begins with a presentation of the *polar decomposition* of the most general Möbius self-transformation of the complex open unit disc $\mathbb{D}$,

$$z \mapsto e^{i\theta} \frac{a + z}{1 + \overline{a}z} = e^{i\theta}(a \oplus z), \tag{1.2}$$

where $a, z \in \mathbb{D}$, $\bar{a}$ being the complex conjugate of a, and where $\theta \in \mathbb{R}$ is a real number [14, p. 211], [19, p. 185], [25, pp. 6 − 7], [35, pp. 177 − 178]. In (1.2) we present the Möbius polar decomposition in a form that suggests the *Möbius addition*, $\oplus$, which we define by the equation

$$a \oplus z = \frac{a + z}{1 + \bar{a}z} \,. \tag{1.3}$$

Naturally, Möbius subtraction, $\ominus$, is given by $a \ominus z = a \oplus (-z)$, so that $z \ominus z = 0$ and $\ominus z = 0 \ominus z = 0 \oplus (-z) = -z$.

Interestingly, Möbius addition possesses the *gyroautomorphic inverse property*

$$\ominus (a \oplus b) = \ominus a \ominus b \tag{1.4}$$

and the *left cancellation law*

$$\ominus a \oplus (a \oplus z) = z \tag{1.5}$$

for all $a, b, z \in \mathbb{D}$. Contrasting the left cancellation law (1.5) of Möbius addition, care must be taken with its right counterpart since, in general,

$$(a \oplus z) \ominus z \neq a \,. \tag{1.6}$$

Indeed, in (1.15) below we will introduce a secondary binary operation, $\boxplus$, which is a binary *cooperation* called a Möbius *coaddition*, and which gives rise to the *right cancellation law*, (1.18), of Möbius addition.

Möbius addition is neither commutative nor associative. Fortunately, the failure of associativity is a source of rich theory, the theory of gyrogroups and gyrovector spaces, as we will see in this book.

To measure the extent of which Möbius addition deviates from commutativity we define the *gyrations*

$$\mathrm{gyr}[a, b] = \frac{a \oplus b}{b \oplus a} = \frac{1 + a\bar{b}}{1 + \bar{a}b} \tag{1.7}$$

for $a, b \in \mathbb{D}$. Being unimodular complex numbers, $|\mathrm{gyr}[a, b]| = 1$, gyrations represent rotations of the disc $\mathbb{D}$ about its center. In the special case when $a = 0$ or $b = 0$ or $a = \lambda b$ for some real number λ, the gyration $\mathrm{gyr}[a, b]$ is trivial, that, is, $\mathrm{gyr}[a, b] = 1$. The number -1 cannot be represented by a gyration since the equation $\mathrm{gyr}[a, b] = -1$ has no solution for a and b in the disc $\mathbb{D}$.

Gyrations respect Möbius addition,

$$\mathrm{gyr}[a, b](c \oplus d) = \mathrm{gyr}[a, b]c \oplus \mathrm{gyr}[a, b]d \tag{1.8}$$

for all $a, b, c, d \in \mathbb{D}$, and the inverse $\mathrm{gyr}^{-1}[a, b]$ of a gyration $\mathrm{gyr}[a, b]$ is the gyration $\mathrm{gyr}[b, a]$,

$$\mathrm{gyr}^{-1}[a, b] = \mathrm{gyr}[b, a] \,, \tag{1.9}$$

as we see from (1.7). Hence, gyrations are special automorphisms of the Möbius groupoid $(\mathbb{D}, \oplus)$. We recall that a groupoid $(G, \oplus)$ is a nonempty set, G, with a binary operation, $\oplus$, and that an automorphism of a groupoid $(G, \oplus)$ is a bijective self map f of G that respects its binary operation $\oplus$, that is, $f(a \oplus b) = f(a) \oplus f(b)$. The set of all automorphisms of a groupoid $(G, \oplus)$ forms a group, denoted $Aut(G, \oplus)$. The formal definition of groups will be presented in Def. 1.3, p. 5.

To emphasize that gyrations are automorphisms, they are also called *gyroautomorphisms*. The automorphism $\phi(z) = -z$ of the disc is not a gyroautomorphism of the disc since the equation $(1 + a\bar{b})/(1 + \bar{a}b) = -1$ for $a, b \in \mathbb{D}$ has no solution in the disc. Consequently, the gyroautomorphisms of the disc do not form a group under gyroautomorphism composition.

It follows from the gyration definition in (1.7) that Möbius addition obeys the *gyrocommutative law*:

$$a \oplus b = \text{gyr}[a, b](b \oplus a) \tag{1.10}$$

for all $a, b \in \mathbb{D}$. The Möbius gyrocommutative law (1.10) is not terribly surprising since it is generated by definition. But, we are not finished!

Coincidentally, the gyration $\text{gyr}[a, b]$, which "repairs" in (1.10) the breakdown of commutativity in Möbius addition, "repairs" the breakdown of associativity in Möbius addition as well. Indeed, it gives rise to the following gyroassociative laws (left and right),

$$a \oplus (b \oplus z) = (a \oplus b) \oplus \text{gyr}[a, b]z\,,$$
$$(a \oplus b) \oplus z = a \oplus (b \oplus \text{gyr}[b, a]z)\,, \tag{1.11}$$

for all $a, b, z \in \mathbb{D}$. Hence, gyrations measure simultaneously the extent to which Möbius addition deviates from both associativity and commutativity.

Möbius gyrations are expressible in terms of Möbius addition. Indeed, solving the first identity in (1.11) for $\text{gyr}[a, b]z$ by means of the Möbius left cancellation law (1.5) we have

$$\text{gyr}[a, b]z = \ominus(a \oplus b) \oplus \{a \oplus (b \oplus z)\}\,. \tag{1.12}$$

Möbius gyrations $\text{gyr}[a, b]$, $a, b \in \mathbb{D}$, thus endow the Möbius disc groupoid $(\mathbb{D}, \oplus)$ with a group-like structure, naturally called a *gyrogroup*. In fact, Möbius gyrations do more than that. They contribute their own rich structure to the gyrogroup. Thus, for instance, they possess the *loop properties* (left and right)

$$\text{gyr}[a \oplus b, b] = \text{gyr}[a, b]\,,$$
$$\text{gyr}[a, b \oplus a] = \text{gyr}[a, b]\,, \tag{1.13}$$

and the nested gyration identity

$$\text{gyr}[b, \ominus\text{gyr}[b, a]a] = \text{gyr}[a, b]\,, \tag{1.14}$$

as one can straightforwardly check.

Möbius addition $\oplus$, as a primary addition in the disc, comes with *Möbius coaddition*, $\boxplus$, as a secondary addition in the disc. Möbius coaddition is defined in terms of Möbius addition and Möbius gyrations by the equation

$$a \boxplus b = a \oplus \text{gyr}[a, \ominus b]b \tag{1.15}$$

for all $a, b \in \mathbb{D}$. It turns out that Möbius coaddition is commutative (but not associative), given by the equation

$$a \boxplus b = \frac{(1 - |a|^2)b + (1 - |b|^2)a}{1 - |a|^2|b|^2} \tag{1.16}$$

for all $a, b \in \mathbb{D}$, as we see from (1.15) and (1.7).

Replacing b by $\ominus b$ in (1.15), and noting that gyr$[a, \ominus b]$ is an automorphism, it follows from (1.15) that

$$a \boxminus b = a \oplus \text{gyr}[a, b](\ominus b) = a \ominus \text{gyr}[a, b]b \,. \tag{1.17}$$

for all $a, b \in \mathbb{D}$, where we use the obvious notation, $a \boxminus b = a \boxplus (\ominus b)$.

In terms of Möbius coaddition, the *right cancellation law* of Möbius addition takes the form

$$(a \oplus b) \boxminus b = a \,. \tag{1.18}$$

The right cancellation law (1.18) is associated with the second right cancellation law

$$(a \boxplus b) \ominus b = a \,, \tag{1.19}$$

as one can straightforwardly check. Identities (1.18) and (1.19) present one of the duality symmetries that Möbius addition and coaddition share.

Guided by the group-like structure of Möbius addition, we define a *gyrogroup* to be a group in which the associative law is replaced by the gyroassociative laws (left and right), and a *gyrocommutative gyrogroup* to be a commutative group in which the commutative and associative laws are replaced by the gyrocommutative and gyroassociative laws. Like Möbius gyrations, the gyrations in the gyrocommutative and gyroassociative laws of an abstract gyrogroup must be automorphisms of the gyrogroup.

Thus, the disc $\mathbb{D}$ with its Möbius addition forms a *gyrocommutative gyrogroup* $(\mathbb{D}, \oplus)$ where, strikingly, the concepts of gyrocommutativity and gyroassociativity of the gyrogroup preserve the flavor of their classical counterparts [54, 55]. Mathematical coincidences are not accidental! Hence, in particular, the gyrations that "coincidentally" repair in (1.10)–(1.11) the breakdown of commutativity and associativity need not be peculiar to Möbius addition, as shown in [15, 16], and as we will see in this book.

Indeed, the generalization of groups into gyrogroups that Möbius addition suggests bears an intriguing resemblance to the generalization of the rational numbers to the real ones. The beginner is initially surprised to discover an irrational number, like $\sqrt{2}$, but soon later he or she is likely to realize that there are more irrational numbers than rational ones. Similarly, the gyrogroup structure of Möbius addition initially comes as a surprise. But, interested explorers may soon realize that there are indications that, in some sense, there are more non-group gyrogroups than groups [15].

1.2   GROUPOIDS, LOOPS, GROUPS, AND GYROGROUPS

In Sec. 1.1, Möbius addition gives rise to gyrations. The mere introduction of gyrations turns groups into gyrogroups. Several definitions culminating in the formal definition of gyrogroups follow.

Definition 1.1.   (Binary Operations, Groupoids, and Groupoid Automorphisms). *A binary operation $+$ in a set S is a function $+ : S \times S \rightarrow S$. We use the notation $a + b$ to denote $+(a, b)$ for any $a, b \in S$. A groupoid $(S, +)$ is a nonempty set, S, with a binary operation, $+$. An automorphism ϕ of a groupoid $(S, +)$ is a bijective (that is, one-to-one onto) self-map of S, $\phi : S \rightarrow S$, which preserves its groupoid operation, that is, $\phi(a + b) = \phi(a) + \phi(b)$ for all $a, b \in S$.*

Groupoids may have identity elements. An identity element of a groupoid $(S, +)$ is an element $0 \in S$ such that $0 + s = s + 0 = s$ for all $s \in S$.

Definition 1.2.   (Loops). *A loop is a groupoid $(S, +)$ with an identity element in which each of the two equations $a + x = b$ and $y + a = b$ for the unknowns x and y possesses a unique solution* [37, 34].

Definition 1.3.   (Groups). *A group is a groupoid $(G, +)$ whose binary operation satisfies the following axioms. In G there is at least one element, 0, called a left identity, satisfying*

(G1) $0 + a = a$

for all $a \in G$. There is an element $0 \in G$ satisfying Axiom (G1) such that for each $a \in G$ there is an element $-a \in G$, called a left inverse of a, satisfying

(G2) $-a + a = 0$.

Moreover, the binary operation obeys the associative law

(G3) $(a + b) + c = a + (b + c)$

for all $a, b, c \in G$.

The binary operation in a given set is known as the set operation. The set of all automorphisms of a groupoid $(S, \oplus)$, denoted $Aut(S, \oplus)$, forms a group with group operation given by bijection composition. The identity automorphism is denoted by I. We say that an automorphism τ is *trivial* if $\tau = I$.

Groups are classified into commutative and noncommutative groups.

Definition 1.4.   (Commutative Groups). *A group $(G, +)$ is commutative if its binary operation obeys the commutative law*

(G6) $a + b = b + a$

for all $a, b \in G$.

Definition 1.5.   (Subgroups). *A subset H of a subgroup $(G, +)$ is a subgroup of G if it is nonempty, and H is closed under group compositions and inverses in G, that is, $x, y \in H$ implies $x + y \in H$ and $-x \in H$.*

Theorem 1.6.   (The Subgroup Criterion). *A subset H of a group G is a subgroup if and only if (i) H is nonempty, and (ii) x, $y \in H$ implies $x - y \in H$.*

For a proof of the Subgroup Criterion see any book on group theory.

A most natural generalization of the group concept, motivated by Möbius addition in the disc $\mathbb{D}$, is the concept of the gyrogroup. Taking the key features of the Möbius groupoid $(\mathbb{D}, \oplus)$ as axioms, and guided by the group axioms, we now define the abstract gyrogroup.

Definition 1.7.   (Gyrogroups). *A groupoid $(G, \oplus)$ is a gyrogroup if its binary operation satisfies the following axioms. In G there is at least one element, 0, called a left identity, satisfying*

(G1) $0 \oplus a = a$

for all $a \in G$. There is an element $0 \in G$ satisfying axiom (G1) such that for each $a \in G$ there is an element $\ominus a \in G$, called a left inverse of a, satisfying

(G2) $\ominus a \oplus a = 0$.

Moreover, for any $a, b, c \in G$ there exists a unique element $\mathrm{gyr}[a, b]c \in G$ such that the binary operation obeys the left gyroassociative law

(G3) $a \oplus (b \oplus c) = (a \oplus b) \oplus \mathrm{gyr}[a, b]c$.

The map $\mathrm{gyr}[a, b] : G \to G$ given by $c \mapsto \mathrm{gyr}[a, b]c$ is an automorphism of the groupoid $(G, \oplus)$, that is,

(G4) $\mathrm{gyr}[a, b] \in Aut(G, \oplus)$,

and the automorphism $\mathrm{gyr}[a, b]$ of G is called the gyroautomorphism, or the gyration, of G generated by $a, b \in G$. The operator $\mathrm{gyr} : G \times G \to Aut(G, \oplus)$ is called the gyrator of G. Finally, the gyroautomorphism $\mathrm{gyr}[a, b]$ generated by any $a, b \in G$ possesses the left loop property

(G5) $\mathrm{gyr}[a, b] = \mathrm{gyr}[a \oplus b, b]$.

Definition 1.8.   (Gyrocommutative Gyrogroups). *A gyrogroup $(G, \oplus)$ is gyrocommutative if its binary operation obeys the gyrocommutative law*

(G6) $a \oplus b = \mathrm{gyr}[a, b](b \oplus a)$

for all $a, b \in G$.

The gyrogroup axioms $(G1)-(G5)$ in Def. 1.7 are classified into three classes:

(1) The first pair of axioms, $(G1)$ and $(G2)$, is a reminiscent of the group axioms.

(2) The last pair of axioms, $(G4)$ and $(G5)$, presents the gyrator axioms.

(3) The middle axiom, $(G3)$, is a hybrid axiom linking the two pairs of axioms in (1) and (2).

As in group theory, we use the notation $a \ominus b = a \oplus (\ominus b)$ in gyrogroup theory as well.

In full analogy with groups, gyrogroups are classified into gyrocommutative and non-gyrocommutative gyrogroups.

Clearly, a (commutative) group is a degenerate (gyrocommutative) gyrogroup whose gyroautomorphisms are all trivial. The algebraic structure of gyrogroups is, accordingly, richer than that of groups. Thus, without losing the flavor of the group structure we have generalized it into the gyrogroup structure to suit the needs of Möbius addition in the disc. Fortunately, the gyrogroup structure is by no means restricted to Möbius addition in the disc. Rather, it abounds in group theory as demonstrated, for instance, in [15] and [16], where finite and infinite gyrogroups, both gyrocommutative and non-gyrocommutative, are presented and studied. Some first gyrogroup theorems, some of which are analogous to group theorems, are presented in [55, Chap. 2], [61, Chap. 2] and in Sec. 1.4.

The gyrogroup operation of any gyrogroup has an associated dual operation, called a *cooperation*, the definition of which follows.

Definition 1.9.   (Gyrogroup Cooperation). Let $(G, \oplus)$ be a gyrogroup with operation $\oplus$. The gyrogroup cooperation $\boxplus$ is a second binary operation in G, given by the equation

$$a \boxplus b = a \oplus \mathrm{gyr}[a, \ominus b]b \tag{1.20}$$

for all $a, b \in G$.

The gyrogroup cooperation $\boxplus$ shares remarkable duality symmetries with its associated gyrogroup operation $\oplus$ as, for instance, the one in Identities 1.49–1.50, p. 15. Furthermore, the gyrogroup cooperation will prove necessary for our mission to capture analogies with classical results. The way the gyrogroup cooperation is used to capture analogies is demonstrated, for instance, by the two right cancellation laws in (1.18)–(1.19).

1.3   MÖBIUS GYROGROUPS: FROM THE DISC TO THE BALL

The gyrocommutative gyrogroup structure in Def. 1.8 is tailor-made for Möbius addition in the disc. However, it suits Möbius addition in the ball of any real inner product space as well. To see this, let us identify complex numbers $u = u_1 + iu_2$ of the complex plane $\mathbb{C}$ with vectors $\mathbf{u} = (u_1, u_2)$ of the Euclidean plane $\mathbb{R}^2$ in the usual way,

$$\mathbb{C} \ni u = u_1 + iu_2 = (u_1, u_2) = \mathbf{u} \in \mathbb{R}^2. \tag{1.21}$$

Then, the equations

$$\bar{u}v + u\bar{v} = 2\mathbf{u} \cdot \mathbf{v},$$

$$|u| = \|\mathbf{u}\| \tag{1.22}$$

give the inner product and the norm in the Euclidean plane $\mathbb{R}^2$, so that Möbius addition in the disc $\mathbb{D}$ of $\mathbb{C}$ becomes Möbius addition in the disc $\mathbb{R}^2_{s=1} = \{\mathbf{v} \in \mathbb{R}^2 : \|\mathbf{v}\| < s = 1\}$ of $\mathbb{R}^2$. Indeed, it follows from (1.3) and (1.22) that

$$
\begin{aligned}
\mathbb{D} \ni u \oplus v &= \frac{u + v}{1 + \bar{u}v} \\
&= \frac{(1 + u\bar{v})(u + v)}{(1 + \bar{u}v)(1 + u\bar{v})} \\
&= \frac{(1 + \bar{u}v + u\bar{v} + |v|^2)u + (1 - |u|^2)v}{1 + \bar{u}v + u\bar{v} + |u|^2|v|^2} \\
&= \frac{(1 + 2\mathbf{u}\cdot\mathbf{v} + \|\mathbf{v}\|^2)\mathbf{u} + (1 - \|\mathbf{u}\|^2)\mathbf{v}}{1 + 2\mathbf{u}\cdot\mathbf{v} + \|\mathbf{u}\|^2\|\mathbf{v}\|^2} \\
&= \mathbf{u} \oplus \mathbf{v} \in \mathbb{R}^2_{s=1}
\end{aligned}
\tag{1.23}
$$

for all $u, v \in \mathbb{D}$ and all $\mathbf{u}, \mathbf{v} \in \mathbb{R}^2_{s=1}$. The last equation in (1.23) is a vector equation, so that its restriction to the ball of the Euclidean two-dimensional space $\mathbb{R}^2$ is a mere artifact. Indeed, it survives unimpaired in higher dimensions, suggesting the following definition of Möbius addition in the ball of any real inner product space.

Definition 1.10.   (Möbius Addition in the Ball). Let $\mathbb{V}$ be a real inner product space [28], and let $\mathbb{V}_s$ be the s-ball of $\mathbb{V}$,

$$
\mathbb{V}_s = \{\mathbb{V}_s \in \mathbb{V} : \|\mathbf{v}\| < s\},
\tag{1.24}
$$

for any fixed $s > 0$. Möbius addition $\oplus$ is a binary operation in $\mathbb{V}_s$ given by the equation

$$
\mathbf{u} \oplus \mathbf{v} = \frac{(1 + \frac{2}{s^2}\mathbf{u}\cdot\mathbf{v} + \frac{1}{s^2}\|\mathbf{v}\|^2)\mathbf{u} + (1 - \frac{1}{s^2}\|\mathbf{u}\|^2)\mathbf{v}}{1 + \frac{2}{s^2}\mathbf{u}\cdot\mathbf{v} + \frac{1}{s^4}\|\mathbf{u}\|^2\|\mathbf{v}\|^2}
\tag{1.25}
$$

where $\cdot$ and $\|\cdot\|$ are the inner product and norm that the ball $\mathbb{V}_s$ inherits from its space $\mathbb{V}$ and where, ambiguously, + denotes both addition of real numbers on the real line $\mathbb{R}$ and addition of vectors in $\mathbb{V}$.

Without loss of generality, one may select $s = 1$ in Def. 1.10. We, however, prefer to keep s as a free positive parameter in order to exhibit the result that in the limit as $s \to \infty$, the ball $\mathbb{V}_s$ expands to the whole of its real inner product space $\mathbb{V}$, and Möbius addition, $\oplus$, in $\mathbb{V}_s$ reduces to vector addition, +, in $\mathbb{V}$.

Being special automorphisms, Möbius gyrations form an important ingredient of the Möbius disc gyrogroup $(\mathbb{D}, \oplus)$. Following the extension of Möbius addition from the disc to the ball, it is desirable to extend Möbius gyrations from the disc to the ball as well. On first glance this task seems impossible, since Möbius gyrations in the disc are given by (1.7) in terms of a division of complex numbers by nonzero complex numbers, an operation that cannot be extended from the disc to the ball. Fortunately, however, identity (1.12) comes to the rescue. This identity expresses gyrations in terms of Möbius addition in the disc, an operation that we have just extended from the disc to the ball in (1.23), as formalized in Def. 1.10.

As suggested by (1.12), and in agreement with Def. 1.7 of gyrogroups, the definition of Möbius gyrations in the ball follows.

Definition 1.11.   (Möbius Gyrations in the Ball). Let $(\mathbb{V}_s, \oplus)$ be a Möbius groupoid as defined in Def. 1.10. The gyrator gyr is the map gyr : $\mathbb{V}_s \times \mathbb{V}_s \rightarrow Aut(\mathbb{V}_s, \oplus)$ given by the equation

$$\text{gyr}[\mathbf{a}, \mathbf{b}]\mathbf{z} = \ominus(\mathbf{a}\oplus\mathbf{b})\oplus\{\mathbf{a}\oplus(\mathbf{b}\oplus\mathbf{z})\} \tag{1.26}$$

for all $\mathbf{a}, \mathbf{b}, \mathbf{z} \in \mathbb{V}_s$. The automorphisms gyr$[\mathbf{a}, \mathbf{b}]$ of $\mathbb{V}_s$ are called gyrations, or gyroautomorphisms, of the ball $\mathbb{V}_s$.

It is anticipated in Def. 1.11 that gyrations of the ball are automorphisms of the ball groupoid $(\mathbb{V}_s, \oplus)$. To see that this is indeed the case, we note that (1.26) can be manipulated by means of Def. 1.10 of Möbius addition, obtaining (with the help of computer algebra)

$$\text{gyr}[\mathbf{u}, \mathbf{v}]\mathbf{w} = \ominus(\mathbf{u}\oplus\mathbf{v})\oplus\{\mathbf{u}\oplus(\mathbf{v}\oplus\mathbf{w})\} = \mathbf{w} + 2\frac{A\mathbf{u} + B\mathbf{v}}{D}, \tag{1.27}$$

where

$$A = -\frac{1}{s^4}\mathbf{u}{\cdot}\mathbf{w}\|\mathbf{v}\|^2 + \frac{1}{s^2}\mathbf{v}{\cdot}\mathbf{w} + \frac{2}{s^4}(\mathbf{u}{\cdot}\mathbf{v})(\mathbf{v}{\cdot}\mathbf{w}),$$

$$B = -\frac{1}{s^4}\mathbf{v}{\cdot}\mathbf{w}\|\mathbf{u}\|^2 - \frac{1}{s^2}\mathbf{u}{\cdot}\mathbf{w}, \tag{1.28}$$

$$D = 1 + \frac{2}{s^2}\mathbf{u}{\cdot}\mathbf{v} + \frac{1}{s^4}\|\mathbf{u}\|^2\|\mathbf{v}\|^2,$$

for all $\mathbf{u}, \mathbf{v}, \mathbf{w} \in \mathbb{V}_s$. Owing to the Cauchy-Schwarz inequality [28, p. 20], $D > 0$ for $\mathbf{u}$ and $\mathbf{v}$ in the ball $\mathbb{V}_s$. Indeed, it follows from the definition of D in (1.28) by the Cauchy-Schwarz inequality that

$$\begin{aligned}
D &= 1 + \frac{2}{s^2}\mathbf{u}{\cdot}\mathbf{v} + \frac{1}{s^4}\|\mathbf{u}\|^2\|\mathbf{v}\|^2 \\
&\geq 1 - \frac{2}{s^2}\|\mathbf{u}\|\|\mathbf{v}\| + \frac{1}{s^4}\|\mathbf{u}\|^2\|\mathbf{v}\|^2 \\
&= \left(1 - \frac{1}{s^2}\|\mathbf{u}\|\|\mathbf{v}\|\right)^2 \\
&> 0
\end{aligned} \tag{1.29}$$

for all $\mathbf{u}, \mathbf{v} \in \mathbb{V}_s$.

Remark 1.12.   *By expanding the domain of $\mathbf{w}$ from the ball $\mathbb{V}_s$ to the space $\mathbb{V}$ in (1.27)–(1.28), we can extend the gyrations gyr$[\mathbf{u}, \mathbf{v}]$ to invertible linear maps of $\mathbb{V}$ onto itself for all $\mathbf{u}, \mathbf{v} \in \mathbb{V}_s$.*

In the special case when (i) $\mathbf{u} = \mathbf{0}$, or (ii) $\mathbf{v} = \mathbf{0}$, or (iii) $\mathbf{u} \| \mathbf{v}$, the gyration $\mathrm{gyr}[\mathbf{u}, \mathbf{v}]$ is trivial, that is, $\mathrm{gyr}[\mathbf{u}, \mathbf{v}]\mathbf{w} = \mathbf{w}$ for all $\mathbf{w} \in \mathbb{V}_s$, as one can readily see from (1.27).

It follows from (1.27)–(1.28) straightforwardly (the use of computer algebra is recommended) that

$$\mathrm{gyr}[\mathbf{v}, \mathbf{u}](\mathrm{gyr}[\mathbf{u}, \mathbf{v}]\mathbf{w}) = \mathbf{w} \qquad (1.30)$$

for all $\mathbf{u}, \mathbf{v}, \mathbf{w} \in \mathbb{V}_s$, so that gyrations of the ball are invertible, the inverse $\mathrm{gyr}^{-1}[\mathbf{u}, \mathbf{v}]$ of $\mathrm{gyr}[\mathbf{u}, \mathbf{v}]$ being $\mathrm{gyr}[\mathbf{v}, \mathbf{u}]$,

$$\mathrm{gyr}^{-1}[\mathbf{u}, \mathbf{v}] = \mathrm{gyr}[\mathbf{v}, \mathbf{u}] . \qquad (1.31)$$

Furthermore, it follows from (1.27)–(1.28) straightforwardly (the use of computer algebra is recommended) that Möbius gyrations of the ball preserve the inner product that the ball $\mathbb{V}_s$ inherits from its real inner product space $\mathbb{V}$, that is,

$$\mathrm{gyr}[\mathbf{u}, \mathbf{v}]\mathbf{a} \cdot \mathrm{gyr}[\mathbf{u}, \mathbf{v}]\mathbf{b} = \mathbf{a} \cdot \mathbf{b} \qquad (1.32)$$

for all $\mathbf{a}, \mathbf{b}, \mathbf{u}, \mathbf{v} \in \mathbb{V}_s$. This implies, in turn, that Möbius gyrations keep invariant the norm that the ball $\mathbb{V}_s$ inherits from its real inner product space $\mathbb{V}$,

$$\|\mathrm{gyr}[\mathbf{u}, \mathbf{v}]\mathbf{a}\| = \|\mathbf{a}\| . \qquad (1.33)$$

As such, Möbius gyrations of the ball preserve Möbius addition in the ball, so that Möbius gyrations are automorphisms of the Möbius ball groupoid $(\mathbb{V}_s, \oplus)$. Being special automorphisms, Möbius gyrations are also called gyroautomorphisms of the ball. The automorphism $\phi(\mathbf{z}) = -\mathbf{z}$ of the ball cannot be represented by a gyroautomorphism of the ball, as verified in [55, p. 70]. Hence, the gyroautomorphisms of the ball do not form a group under gyroautomorphism composition.

Not unexpectedly, Möbius addition $\oplus$ in the ball $\mathbb{V}_s$ of any real inner product space $\mathbb{V}$ preserves the structure it has in the disc. It thus gives rise to the Möbius gyrocommutative gyrogroup $(\mathbb{V}_s, \oplus)$ in the ball, studied in [55]. Möbius addition (1.25) in the ball is known in the literature as a *hyperbolic translation* [2, 40]. However, its gyrogroup structure went unnoticed until it was uncovered in 1988 [52, 53] in the context of Einstein velocity addition of special relativity theory. Indeed, like Möbius addition, Einstein velocity addition law of special relativity gives rise to the Einstein gyrocommutative gyrogroup $(\mathbb{V}_s, \oplus_{_E})$ in the ball [52, 54, 56, 61], which we will study in Sec. 2.3.

The evolution of Möbius addition and gyrations in the disc does not stop at the level of gyrogroups. It continues into the regime of *gyrovector spaces*. Strikingly, gyrovector spaces form the setting for hyperbolic geometry just as vector spaces form the setting for Euclidean geometry. In particular, Möbius gyrovector spaces form the setting for the Poincaré ball model of hyperbolic geometry while, similarly, Einstein gyrovector spaces form the setting for the Beltrami-Klein ball model of hyperbolic geometry [54, 55, 57, 58, 61, 65]. Accordingly, a gyrovector space approach to analytic hyperbolic geometry, fully analogous to the common vector space approach to Euclidean geometry [22], is developed in [55, 61] and in this book.

Equations (1.2)–(1.14) of Möbius addition and gyrations in the disc remain valid in the ball as well. Hence, they clearly pass the tests of simplicity, beauty, and most importantly, generality.

Accordingly, the far-reaching evolution of Möbius addition and gyrations in the disc into the abstract gyrogroup is likely to set new standards in twenty-first century hyperbolic geometry, special relativity theory, and nonassociative algebra. The symbiosis between these three theories vastly enriches them, largely due to the presence of gyrations.

1.4   FIRST GYROGROUP THEOREMS

In this section we present several first gyrogroup theorems. These, and more advanced gyrogroup theorems, are found in [54, 55, 61].

While it is clear how to define a right identity and a right inverse in a gyrogroup, the existence of such elements is not presumed. Indeed, the existence of a unique identity and a unique inverse, both left and right, is a consequence of the gyrogroup axioms, as the following theorem shows, along with other immediate results.

Theorem 1.13.   *Let $(G, +)$ be a gyrogroup. For any elements $a, b, c, x \in G$ we have:*

(1) If $a + b = a + c$, then $b = c$ (general left cancellation law; see item (9) below).

(2) $\mathrm{gyr}[0, a] = I$ for any left identity 0 in G.

(3) $\mathrm{gyr}[x, a] = I$ for any left inverse x of a in G.

(4) $\mathrm{gyr}[a, a] = I$

(5) There is a left identity which is a right identity.

(6) There is only one left identity.

(7) Every left inverse is a right inverse.

(8) There is only one left inverse, $-a$, of a, and $-(-a) = a$.

(9) $-a + (a + b) = b$ (Left Cancellation Law).

(10) $\mathrm{gyr}[a, b]x = -(a + b) + \{a + (b + x)\}$ (The Gyrator Identity).

(11) $\mathrm{gyr}[a, b]0 = 0$.

(12) $\mathrm{gyr}[a, b](-x) = -\mathrm{gyr}[a, b]x$.

(13) $\mathrm{gyr}[a, 0] = I$.

Proof.

(1) Let x be a left inverse of a corresponding to a left identity, 0, in G. We have $x + (a + b) = x + (a + c)$. By left gyroassociativity, $(x + a) + \mathrm{gyr}[x, a]b = (x + a) + \mathrm{gyr}[x, a]c$. Since 0 is a left identity, $\mathrm{gyr}[x, a]b = \mathrm{gyr}[x, a]c$. Since automorphisms are bijective, $b = c$.

(2) By left gyroassociativity we have for any left identity 0 of G, $a + x = 0 + (a + x) = (0 + a) + \mathrm{gyr}[0, a]x = a + \mathrm{gyr}[0, a]x$. Hence, by (1) above we have $x = \mathrm{gyr}[0, a]x$ for all $x \in G$ so that $\mathrm{gyr}[0, a] = I$.

(3) By the left loop property and by (2) above we have gyr$[x, a]$ = gyr$[x + a, a]$ = gyr$[0, a]$ = I.

(4) Follows from an application of the left loop property and (2) above.

(5) Let x be a left inverse of a corresponding to a left identity, 0, of G. Then by left gyroassociativity and (3) above, $x + (a + 0) = (x + a) + $ gyr$[x, a]0 = 0 + 0 = 0 = x + a$. Hence, by (1), $a + 0 = a$ for all $a \in G$ so that 0 is a right identity.

(6) Suppose 0 and 0^* are two left identities, one of which, say 0, is also a right identity. Then $0 = 0^* + 0 = 0^*$.

(7) Let x be a left inverse of a. Then $x + (a + x) = (x + a) + $ gyr$[x, a]x = 0 + x = x = x + 0$, by left gyroassociativity, (G2) of Def. 1.7, (3), and (5) and (6) above. By (1) we have $a + x = 0$ so that x is a right inverse of a.

(8) Suppose x and y are left inverses of a. By (7) above, they are also right inverses, so $a + x = 0 = a + y$. By (1), $x = y$. Let $-a$ be the resulting unique inverse of a. Then $-a + a = 0$ so that the inverse $-(-a)$ of $-a$ is a.

(9) By left gyroassociativity and by (3) above we have $-a + (a + b) = (-a + a) + $ gyr$[-a, a]b = b$.

(10) By an application of the left cancellation law (9) to the left gyroassociative law (G3) in Def. 1.7 we obtain (10).

(11) We obtain (11) from (10) with $x = 0$.

(12) Since gyr$[a, b]$ is an automorphism of $(G, +)$ we have from (11) gyr$[a, b](-x) + $ gyr$[a, b]x = $ gyr$[a, b](-x + x) = $ gyr$[a, b]0 = 0$, and hence the result.

(13) We obtain (13) from (10) with $b = 0$, and a left cancellation, (9).   □

Möbius gyrogroups possess the gyroautomorphic inverse property (1.4). In general, however, $-(a + b) \neq -a - b$ in a gyrogroup $(G, +)$, as we will see in Theorem 2.2 of Chap. 2, p. 33. Hence, the following theorem proves useful.

Theorem 1.14.   (Gyrosum Inversion Law). *For any two elements a, b of a gyrogroup $(G, +)$ we have the gyrosum inversion law*

$$- (a + b) = \text{gyr}[a, b](-b - a). \tag{1.34}$$

Proof. By the gyrator identity in Theorem 1.13(10) and a left cancellation, Theorem 1.13(9), we have

$$\begin{aligned} \text{gyr}[a, b](-b - a) &= -(a + b) + (a + (b + (-b - a))) \\ &= -(a + b) + (a - a) \\ &= -(a + b). \end{aligned} \tag{1.35}$$

□

As a special case of the gyrosum inversion law (1.34) we note that

$$- (a + a) = -a - a \tag{1.36}$$

in any gyrogroup $(G, +)$. We call (1.36) a *restricted gyroautomorphic inverse property*. Restricted gyroautomorphic inverse properties result from the gyrosum inversion (1.34) whenever gyr$[a, b]$ is trivial, gyr$[a, b] = I$.

Lemma 1.15. *For any two elements, a and b, of a gyrogroup $(G, +)$, we have*

$$\begin{aligned}
\text{gyr}[a, b]b &= -\{-(a + b) + a\}, \\
\text{gyr}[a, -b]b &= -(a - b) + a.
\end{aligned} \tag{1.37}$$

Proof. The first identity in (1.37) follows from Theorem 1.13(10) with $x = -b$, and Theorem 1.13(12), and the second part of Theorem 1.13(8). The second identity in (1.37) follows from the first one by replacing b by $-b$.   □

A nested gyroautomorphism is a gyration generated by points that depend on another gyration as, for instance, (1.14) and some gyrations in (1.38)–(1.40) below.

Theorem 1.16. *Any three elements a, b, c of a gyrogroup $(G, +)$ satisfy the nested gyroautomorphism identities*

$$\text{gyr}[a, b + c]\text{gyr}[b, c] = \text{gyr}[a + b, \text{gyr}[a, b]c]\text{gyr}[a, b], \tag{1.38}$$

$$\text{gyr}[a + b, -\text{gyr}[a, b]b]\text{gyr}[a, b] = I, \tag{1.39}$$

$$\text{gyr}[a, -\text{gyr}[a, b]b]\text{gyr}[a, b] = I, \tag{1.40}$$

and the gyroautomorphism product identities

$$\text{gyr}[-a, a + b]\text{gyr}[a, b] = I, \tag{1.41}$$

$$\text{gyr}[b, a + b]\text{gyr}[a, b] = I. \tag{1.42}$$

Proof. By two successive applications of the left gyroassociative law in two different ways, we obtain the following two chains of equations for all $a, b, c, x \in G$,

$$\begin{aligned}
a + (b + (c + x)) &= a + ((b + c) + \text{gyr}[b, c]x) \\
&= (a + (b + c)) + \text{gyr}[a, b + c]\text{gyr}[b, c]x,
\end{aligned} \tag{1.43}$$

and

$$\begin{aligned}
a + (b + (c + x)) &= (a + b) + \mathrm{gyr}[a, b](c + x) \\
&= (a + b) + (\mathrm{gyr}[a, b]c + \mathrm{gyr}[a, b]x) \\
&= ((a + b) + \mathrm{gyr}[a, b]c) + \mathrm{gyr}[a + b, \mathrm{gyr}[a, b]c]\mathrm{gyr}[a, b]x \\
&= (a + (b + c)) + \mathrm{gyr}[a + b, \mathrm{gyr}[a, b]c]\mathrm{gyr}[a, b]x \, .
\end{aligned} \tag{1.44}$$

By comparing the extreme right-hand sides of these two chains of equations, and by employing the left cancellation law, Theorem 1.13(1), we obtain the identity

$$\mathrm{gyr}[a, b + c]\mathrm{gyr}[b, c]x = \mathrm{gyr}[a + b, \mathrm{gyr}[a, b]c]\mathrm{gyr}[a, b]x \tag{1.45}$$

for all $x \in G$, thus verifying (1.38).

In the special case when $c = -b$, (1.38) reduces to (1.39), noting that the left-hand side of (1.38) becomes trivial owing to items (2) and (3) of Theorem 1.13.

Identity (1.40) results from the following chain of equations, which are numbered for subsequent derivation:

$$\begin{aligned}
I &\overset{(1)}{=\!=} \mathrm{gyr}[a + b, -\mathrm{gyr}[a, b]b]\mathrm{gyr}[a, b] \\
&\overset{(2)}{=\!=} \mathrm{gyr}[(a + b) - \mathrm{gyr}[a, b]b, -\mathrm{gyr}[a, b]b]\mathrm{gyr}[a, b] \\
&\overset{(3)}{=\!=} \mathrm{gyr}[a + (b - b), -\mathrm{gyr}[a, b]b]\mathrm{gyr}[a, b] \\
&\overset{(4)}{=\!=} \mathrm{gyr}[a, -\mathrm{gyr}[a, b]b]\mathrm{gyr}[a, b] \, .
\end{aligned} \tag{1.46}$$

Derivation of the numbered equalities in (3.68) follows.

(1) Follows from (1.39).

(2) Follows from (1) by the left loop property.

(3) Follows from (2) by the left gyroassociative law. Indeed, an application of the left gyroassociative law to the first entry of the left gyration in (3) gives the first entry of the left gyration in (2), that is, $a + (b - b) = (a + b) - \mathrm{gyr}[a, b]b$.

(4) Follows from (3) immediately.

To verify (1.41) we consider the special case of (1.38) when $b = -a$, obtaining

$$\begin{aligned}
\mathrm{gyr}[a, -a + c]\mathrm{gyr}[-a, c] &= \mathrm{gyr}[0, \mathrm{gyr}[a, -a]c]\mathrm{gyr}[a, -a] \\
&= I \, ,
\end{aligned} \tag{1.47}$$

where the second identity in (1.47) follows from items (2) and (3) of Theorem 1.13. Replacing a by $-a$ and c by b in (1.47) we obtain (1.41).

Finally, (1.42) is derived from (1.41) by an application of the left loop property to the first gyroautomorphism in (1.41) followed by a left cancellation, Theorem 1.13(9),

$$
\begin{aligned}
I &= \mathrm{gyr}[-a, a + b]\mathrm{gyr}[a, b] \\
&= \mathrm{gyr}[-a + (a + b), a + b]\mathrm{gyr}[a, b] \\
&= \mathrm{gyr}[b, a + b]\mathrm{gyr}[a, b] \,.
\end{aligned}
\tag{1.48}
$$

□

The nested gyroautomorphism identity (1.40) in Theorem 1.16 allows the equation that defines the coaddition $\boxplus$ to be dualized with its corresponding equation in which the roles of the binary operations $\boxplus$ and $\oplus$ are interchanged, as shown in the following theorem.

Theorem 1.17.   *Let $(G, \oplus)$ be a gyrogroup with cooperation $\boxplus$ given in Def. 1.9, p. 7, by the equation*

$$
a \boxplus b = a \oplus \mathrm{gyr}[a, \ominus b]b \,.
\tag{1.49}
$$

Then

$$
a \oplus b = a \boxplus \mathrm{gyr}[a, b]b \,.
\tag{1.50}
$$

Proof.   Let a and b be any two elements of G. By (1.49) and (1.40) we have

$$
\begin{aligned}
a \boxplus \mathrm{gyr}[a, b]b &= a \oplus \mathrm{gyr}[a, \ominus \mathrm{gyr}[a, b]b]\mathrm{gyr}[a, b]b \\
&= a \oplus b \,,
\end{aligned}
\tag{1.51}
$$

thus verifying (1.50).

□

In view of the duality symmetry that Identities (1.49) and (1.50) share, the gyroautomorphisms $\mathrm{gyr}[a, b]$ and $\mathrm{gyr}[a, \ominus b]$ may be considered dual to each other.

We naturally use the notation

$$
a \boxminus b = a \boxplus (\ominus b) \,,
\tag{1.52}
$$

in a gyrogroup $(G, \oplus)$, so that, by (1.52), (1.49) and Theorem 1.13(12),

$$
\begin{aligned}
a \boxminus b &= a \boxplus (\ominus b) \\
&= a \oplus \mathrm{gyr}[a, b](\ominus b) \\
&= a \ominus \mathrm{gyr}[a, b]b \,,
\end{aligned}
\tag{1.53}
$$

and, hence,

$$
a \boxminus a = a \ominus a = 0 \,,
\tag{1.54}
$$

as it should. Identity (1.54), in turn, implies the equality between the inverses of $a \in G$ with respect to $\oplus$ and $\boxplus$,

$$
\boxminus a = \ominus a
\tag{1.55}
$$

for all $a \in G$.

As an application of the left cancellation law in Theorem 1.13(9) we present in the next theorem the gyrogroup counterpart (1.56) of the simple, but important, group identity $(-a + b) + (-b + c) = -a + c$.

Theorem 1.18.   *Let $(G, +)$ be a gyrogroup. Then*

$$(-a + b) + \text{gyr}[-a, b](-b + c) = -a + c \tag{1.56}$$

for all $a, b, c \in G$.

Proof. By the left gyroassociative law and the left cancellation law, and using the notation $d = -b + c$, we have,

$$\begin{aligned}
(-a + b) + \text{gyr}[-a, b](-b + c) &= (-a + b) + \text{gyr}[-a, b]d \\
&= -a + (b + d) \\
&= -a + (b + (-b + c)) \\
&= -a + c.
\end{aligned} \tag{1.57}$$

$\square$

Theorem 1.19.   **(The Gyrotranslation Theorem, I).** *Let $(G, +)$ be a gyrogroup. Then*

$$-(-a + b) + (-a + c) = \text{gyr}[-a, b](-b + c) \tag{1.58}$$

for all $a, b, c \in G$.

Proof. Identity (1.58) is a rearrangement of Identity (1.56) obtained by a left cancellation. $\square$

The importance of Identity (1.58) lies in the analogy it shares with its group counterpart, $-(-a + b) + (-a + c) = -b + c$.

The identity of Theorem 1.18 can readily be generalized to any number of terms, for instance,

$$(-a + b) + \text{gyr}[-a, b]\{(-b + c) + \text{gyr}[-b, c](-c + d)\} = -a + d, \tag{1.59}$$

which generalizes the obvious group identity $(-a + b) + (-b + c) + (-c + d) = -a + d$.

1.5   THE TWO BASIC EQUATIONS OF GYROGROUPS

The two basic equations of gyrogroup theory are

$$a \oplus x = b \tag{1.60}$$

and

$$x \oplus a = b, \tag{1.61}$$

$a, b, x \in G$, for the unknown x in a gyrogroup $(G, \oplus)$.

Let x be a solution of the first basic equation, (1.60). Then we have by (1.60) and the left cancellation law, Theorem 1.13(9),

$$\ominus a \oplus b = \ominus a \oplus (a \oplus x) = x. \tag{1.62}$$

Hence, if a solution x of (1.60) exists then it must be given by $x = \ominus a \oplus b$, as we see from (1.62).

Conversely, $x = \ominus a \oplus b$ is, indeed, a solution of (1.60) as wee see by substituting $x = \ominus a \oplus b$ into (1.60) and applying the left cancellation law in Theorem 1.13(9). Hence, the gyrogroup equation (1.60) possesses the unique solution $x = \ominus a \oplus b$.

The solution of the second basic gyrogroup equation, (1.61), is quiet different from that of the first, (1.60), owing to the noncommutativity of the gyrogroup operation. Let x be a solution of (1.61). Then we have the following chain of equations, which are numbered for subsequent derivation:

$$
\begin{aligned}
x &\overset{(1)}{=\!=} x \oplus 0 \\
&\overset{(2)}{=\!=} x \oplus (a \ominus a) \\
&\overset{(3)}{=\!=} (x \oplus a) \oplus \mathrm{gyr}[x, a](\ominus a) \\
&\overset{(4)}{=\!=} (x \oplus a) \ominus \mathrm{gyr}[x, a]a \\
&\overset{(5)}{=\!=} (x \oplus a) \ominus \mathrm{gyr}[x \oplus a, a]a \\
&\overset{(6)}{=\!=} b \ominus \mathrm{gyr}[b, a]a \overset{(7)}{=\!=} b \boxminus a.
\end{aligned}
\tag{1.63}
$$

Derivation of the numbered equalities in (1.63) follows.

(1) Follows from the existence of a unique identity element, 0, in the gyrogroup $(G, \oplus)$ by Theorem 1.13.

(2) Follows from the existence of a unique inverse element $\ominus a$ of a in the gyrogroup $(G, \oplus)$ by Theorem 1.13.

(3) Follows from (2) by the left gyroassociative law in Axiom (G3) of gyrogroups in Def. 1.7.

(4) Follows from (3) by Theorem 1.13(12).

(5) Follows from (4) by the left loop property (G5) of gyrogroups in Def. 1.7.

(6) Follows from (5) by the assumption that x is a solution of (1.61).

(7) Follows from (6) by (1.53).

Hence, if a solution x of (1.61) exists then it must be given by $x = b \boxminus a$, as we see from (1.63).

Conversely, $x = b \boxminus a$ is, indeed, a solution of (1.61), as we see from the following chain of equations:

$$
\begin{aligned}
x \oplus a &\overset{(1)}{=\!=\!=} (b \boxminus a) \oplus a \\
&\overset{(2)}{=\!=\!=} (b \ominus \mathrm{gyr}[b, a]a) \oplus a \\
&\overset{(3)}{=\!=\!=} (b \ominus \mathrm{gyr}[b, a]a) \oplus \mathrm{gyr}[b, \ominus \mathrm{gyr}[b, a]]\mathrm{gyr}[b, a]a \\
&\overset{(4)}{=\!=\!=} b \oplus (\ominus \mathrm{gyr}[b, a]a \oplus \mathrm{gyr}[b, a]a) \\
&\overset{(5)}{=\!=\!=} b \oplus 0 \\
&\overset{(6)}{=\!=\!=} b\,.
\end{aligned}
\tag{1.64}
$$

Derivation of the numbered equalities in (1.64) follows.

(1) Follows from the assumption that $x = b \boxminus a$.

(2) Follows from (1) by (1.53).

(3) Follows from (2) by Identity (1.40) of Theorem 1.16, according to which the gyration product applied to a in (3) is trivial.

(4) Follows from (3) by the left gyroassociative law. Indeed, an application of the left gyroassociative law to (4) results in (3).

(5) Follows from (4) since $\ominus \mathrm{gyr}[b, a]a$ is the unique inverse of $\mathrm{gyr}[b, a]a$.

(6) Follows from (5) since 0 is the unique identity element of the gyrogroup $(G, \oplus)$.

Formalizing the results of this section, we have the following theorem.

Theorem 1.20.   (The Two Basic Equations Theorem). *Let $(G, \oplus)$ be a gyrogroup, and let $a, b \in G$. The unique solution of the equation*

$$
a \oplus x = b
\tag{1.65}
$$

in G for the unknown x is

$$
x = \ominus a \oplus b\,,
\tag{1.66}
$$

and the unique solution of the equation

$$x \oplus a = b \tag{1.67}$$

in G for the unknown x is

$$x = b \boxminus a \,. \tag{1.68}$$

Owing to Theorem 1.20, gyrogroups are special loops, as we see from Def. 1.2, p. 5. The validity of the solution (1.68) to equation (1.67) follows from the left loop property, as we see from the derivation of equality (5) of the chain of equations (1.63). Accordingly, it is owing to the loop property of gyrogroups that gyrogroups are loops. The term *"loop property"* that was coined in the definition of gyrogroups to describe Axiom (G5) is now justified.

Let $(G, \oplus)$ be a gyrogroup, and let $a \in G$. The maps λ_a and ρ_a of G, given by

$$\begin{aligned} \lambda_a &: \ G \ \to \ G, & \lambda_a &: \ g \ \mapsto \ a \oplus g \,, \\ \rho_a &: \ G \ \to \ G, & \rho_a &: \ g \ \mapsto \ g \oplus a \,, \end{aligned} \tag{1.69}$$

are called, respectively, a *left gyrotranslation* of G by a and a *right gyrotranslation* of G by a. Theorem 1.20 asserts that each of these transformations of G is bijective, that is, it maps G onto itself in a one-to-one manner.

1.6 THE BASIC CANCELLATION LAWS OF GYROGROUPS

The basic cancellation laws of gyrogroup theory are obtained in this section from the basic equations of gyrogroups solved in Sec. 1.5. Substituting the solution (1.66) into its equation (1.65) we obtain the left cancellation law

$$a \oplus (\ominus a \oplus b) = b \tag{1.70}$$

for all $a, b \in G$, already verified in Theorem 1.13(9).

Similarly, substituting the solution (1.68) into its equation (1.67) we obtain the first right cancellation law

$$(b \boxminus a) \oplus a = b \,, \tag{1.71}$$

for all $a, b \in G$. The latter can be dualized, obtaining the second right cancellation law

$$(b \ominus a) \boxplus a = b \,, \tag{1.72}$$

for all $a, b \in G$. Indeed, (1.72) results from the following chain of equations

$$\begin{aligned} b &= b \oplus 0 \\ &= b \oplus (\ominus a \oplus a) \\ &= (b \ominus a) \oplus \text{gyr}[b, \ominus a] a \\ &= (b \ominus a) \oplus \text{gyr}[b \ominus a, \ominus a] a \\ &= (b \ominus a) \boxplus a \,, \end{aligned} \tag{1.73}$$

where we employ the left gyroassociative law, the left loop property, and the definition of the gyrogroup cooperation.

Identities $(1.70)-(1.72)$ form the three basic cancellation laws of gyrogroup theory. Indeed, these cancellation laws are used frequently in the study of gyrogroups and gyrovector spaces.

1.7   COMMUTING AUTOMORPHISMS WITH GYROAUTO-MORPHISMS

The commutativity between automorphisms and gyroautomorphisms of a gyrogroup is not the ordinary one but, rather, it is a special commutative law. In this section we will find that automorphisms of a gyrogroup commute with its gyroautomorphisms in a special way that proves useful in the study of gyrogroups.

Theorem 1.21.   *For any two elements a, b of a gyrogroup $(G, +)$ and any automorphism A of $(G, +)$, $A \in Aut(G, +)$,*

$$A\mathrm{gyr}[a, b] = \mathrm{gyr}[Aa, Ab]A. \tag{1.74}$$

Proof.   For any three elements $a, b, x \in (G, +)$ and any automorphism $A \in Aut(G, +)$ we have by the left gyroassociative law,

$$
\begin{aligned}
(Aa + Ab) + A\mathrm{gyr}[a, b]x &= A((a + b) + \mathrm{gyr}[a, b]x) \\
&= A(a + (b + x)) \\
&= Aa + (Ab + Ax) \\
&= (Aa + Ab) + \mathrm{gyr}[Aa, Ab]Ax.
\end{aligned}
\tag{1.75}
$$

Hence, by a left cancellation, Theorem 1.13(1),

$$A\mathrm{gyr}[a, b]x = \mathrm{gyr}[Aa, Ab]Ax,$$

for all $x \in G$, implying (1.74).   □

As an application of Theorem 1.21 we have the following theorem.

Theorem 1.22.   *Let a, b be any two elements of a gyrogroup $(G, +)$ and let $A \in Aut(G)$ be an automorphism of G. Then*

$$\mathrm{gyr}[a, b] = \mathrm{gyr}[Aa, Ab] \tag{1.76}$$

if and only if the automorphisms A and $\mathrm{gyr}[a, b]$ are commutative.

Proof.   If $\mathrm{gyr}[Aa, Ab] = \mathrm{gyr}[a, b]$ then by Theorem 1.21 the automorphisms $\mathrm{gyr}[a, b]$ and A commute, $A\mathrm{gyr}[a, b] = \mathrm{gyr}[Aa, Ab]A = \mathrm{gyr}[a, b]A$. Conversely, if $\mathrm{gyr}[a, b]$ and A commute then by Theorem 1.21 $\mathrm{gyr}[Aa, Ab] = A\mathrm{gyr}[a, b]A^{-1} = \mathrm{gyr}[a, b]AA^{-1} = \mathrm{gyr}[a, b]$.   □

As a simple, but useful, consequence of Theorem 1.22 we note the elegant identity

$$\mathrm{gyr}[\mathrm{gyr}[a, b]a, \mathrm{gyr}[a, b]b] = \mathrm{gyr}[a, b]. \tag{1.77}$$

1.8   THE GYROSEMIDIRECT PRODUCT

The gyrosemidirect product is a natural generalization of the notion of the semidirect product of group theory.

Definition 1.23.   (Gyroautomorphism Groups, Gyrosemidirect Product). *Let $G = (G, +)$ be a gyrogroup, and let $Aut(G) = Aut(G, +)$ be the automorphism group of G. A gyroautomorphism group, $Aut_0(G)$, of G is any subgroup of $Aut(G)$ containing all the gyroautomorphisms $\text{gyr}[a, b]$ of G, $a, b \in G$. The gyrosemidirect product group*

$$G \times Aut_0(G) \tag{1.78}$$

of a gyrogroup G and any gyroautomorphism group, $Aut_0(G)$ of G, is a group of pairs (x, X), where $x \in G$ and $X \in Aut_0(G)$, with operation given by the gyrosemidirect product

$$(x, X)(y, Y) = (x + Xy, \text{gyr}[x, Xy]XY). \tag{1.79}$$

In analogy with the notion of the semidirect product in group theory, the gyrosemidirect product group

$$G \times Aut(G) \tag{1.80}$$

is called the gyroholomorph of G.

It is anticipated in Def. 1.23 that the gyrosemidirect product set (1.78) of a gyrogroup and any one of its gyroautomorphism groups is a set that forms a group with group operation given by the gyrosemidirect product (1.79). The following theorem shows that this is indeed the case.

Theorem 1.24. *Let $(G, +)$ be a gyrogroup, and let $Aut_0(G, +)$ be a gyroautomorphism group of G. Then the gyrosemidirect product $G \times Aut_0(G)$ is a group, with group operation given by the gyrosemidirect product (1.79).*

Proof. We will show that the set $G \times Aut_0(G)$ with its binary operation (1.79) satisfies the group axioms.
(i) Existence of a left identity: A left identity element of $G \times Aut_0(G)$ is the pair $(0, I)$, where $0 \in G$ is the identity element of G, and $I \in Aut_0(G)$ is the identity automorphism of G. Indeed,

$$(0, I)(a, A) = (0 + Ia, \text{gyr}[0, Ia]IA) = (a, A), \tag{1.81}$$

noting that the gyration in (1.81) is trivial by Theorem 1.13(2).
(ii) Existence of a left inverse: Let $A^{-1} \in Aut_0(G)$ be the inverse automorphism of $A \in Aut_0(G)$. Then, by the gyrosemidirect product (1.79) we have

$$(-A^{-1}a, A^{-1})(a, A) = (-A^{-1}a + A^{-1}a, \text{gyr}[-A^{-1}a, A^{-1}a]A^{-1}A) = (0, I). \tag{1.82}$$

Hence, a left inverse of $(a, A) \in G \times Aut_0(G)$ is the pair $(-A^{-1}a, A^{-1})$,

$$(a, A)^{-1} = (-A^{-1}a, A^{-1}) . \tag{1.83}$$

(iii) Validity of the associative law: We have to show that the successive products in (1.84) and in (1.85) below are equal.

On the one hand, we have

$$(a_1, A_1)((a_2, A_2)(a_3, A_3))$$
$$= (a_1, A_1)(a_2 + A_2 a_3, \text{gyr}[a_2, A_2 a_3]A_2 A_3)$$
$$= (a_1 + A_1(a_2 + A_2 a_3), \text{gyr}[a_1, A_1(a_2 + A_2 a_3)]A_1 \text{gyr}[a_2, A_2 a_3]A_2 A_3) \tag{1.84}$$
$$= (a_1 + (A_1 a_2 + A_1 A_2 a_3), \text{gyr}[a_1, A_1 a_2 + A_1 A_2 a_3]\text{gyr}[A_1 a_2, A_1 A_2 a_3]A_1 A_2 A_3)$$

where we employ the gyrosemidirect product (1.79) and the commuting law (1.74).

On the other hand, we have

$$((a_1, A_1)(a_2, A_2))(a_3, A_3)$$
$$= (a_1 + A_1 a_2, \text{gyr}[a_1, A_1 a_2]A_1 A_2)(a_3, A_3)$$
$$= ((a_1 + A_1 a_2) + \text{gyr}[a_1, A_1 a_2]A_1 A_2 a_3, \tag{1.85}$$
$$\text{gyr}[a_1 + A_1 a_2, \text{gyr}[a_1, A_1 a_2]A_1 A_2 a_3]\text{gyr}[a_1, A_1 a_2]A_1 A_2 A_3)$$

where we employ (1.79).

In order to show that the gyrosemidirect products in (1.84) and (1.85) are equal, using the notation

$$a_1 = a$$
$$A_1 a_2 = b \tag{1.86}$$
$$A_1 A_2 a_3 = c ,$$

we have to establish the identity

$$(a + (b + c), \text{gyr}[a, b + c]\text{gyr}[b, c]A_1 A_2 A_3)$$
$$= ((a + b) + \text{gyr}[a, b]c, \text{gyr}[a + b, \text{gyr}[a, b]c]\text{gyr}[a, b]A_1 A_2 A_3) . \tag{1.87}$$

This identity between two pairs is equivalent to the two identities between their corresponding entries,

$$a + (b + c) = (a + b) + \text{gyr}[a, b]c$$
$$\text{gyr}[a, b + c]\text{gyr}[b, c] = \text{gyr}[a + b, \text{gyr}[a, b]c]\text{gyr}[a, b] . \tag{1.88}$$

The first identity is valid, being the left gyroassociative law, and the second identity is valid by (1.38), p. 13.   □

Instructively, a second proof of Theorem 1.24 is given below.

Proof. A one-to-one map of a set Q_1 onto a set Q_2 is said to be bijective and, accordingly, the map is called a bijection. The set of all bijections of a set Q onto itself forms a group under bijection composition. Let S be the group of all bijections of the set G onto itself under bijection composition. Let each element

$$(a, A) \in S_0 := G \times Aut_0(G) \tag{1.89}$$

act bijectively on the gyrogroup $(G, +)$ according to the equation

$$(a, A)g = a + Ag, \tag{1.90}$$

the unique inverse of (a, A) in $S_0 = G \times Aut_0(G)$ being, by (1.83),

$$(a, A)^{-1} = (-A^{-1}a, A^{-1}). \tag{1.91}$$

Being a set of special bijections of G onto itself, given by (1.90), S_0 is a subset of the group S, $S_0 \subset S$. Employing the subgroup criterion in Theorem 1.6, p. 6, we will show that, under bijection composition, S_0 is a subgroup of the group S.

Two successive bijections $(a, A), (b, B) \in S_0$ of G are equivalent to a single bijection $(c, C) \in S_0$ according to the following chain of equations. Employing successively the bijection (1.90) along with the left gyroassociative law we have

$$\begin{aligned}
(a, A)(b, B)g &= (a, A)(b + Bg) \\
&= a + A(b + Bg) \\
&= a + (Ab + ABg) \\
&= (a + Ab) + \text{gyr}[a, Ab]ABg \\
&= (a + Ab, \text{gyr}[a, Ab]AB)g \\
&=: (c, C)g
\end{aligned} \tag{1.92}$$

for all $g \in G$, $(a, A), (b, B) \in S_0$.

It follows from (1.92) that bijection composition in S_0 is given by the gyrosemidirect product, (1.79),

$$(a, A)(b, B) = (a + Ab, \text{gyr}[a, Ab]AB) \tag{1.93}$$

Finally, for any $(a, A), (b, B) \in S_0$ we have by (1.91) and (1.93),

$$\begin{aligned}
(a, A)(b, B)^{-1} &= (a, A)(-B^{-1}b, B^{-1}) \\
&= (a - AB^{-1}b, \text{gyr}[a, -AB^{-1}b]AB^{-1}) \\
&\in S_0.
\end{aligned} \tag{1.94}$$

Hence, by the subgroup criterion in Theorem 1.6, p. 6, the subset S_0 of the group S of all bijections of G onto itself is a subgroup under bijection composition. But, bijection composition in S_0 is given by the gyrosemidirect product (1.93). Hence, as desired, the set $S_0 = G \times Aut_0(G)$ with composition given by the gyrosemidirect product (1.93) forms a group.   □

The gyrosemidirect product group enables problems in gyrogroups to be converted to the group setting thus gaining access to the powerful group theoretic techniques. Illustrative examples for the use of gyrosemidirect product groups are provided by the proof of the following two Theorems 1.25 and 1.26.

Theorem 1.25.   *Let $(G, +)$ be a gyrogroup, let $a, b \in G$ be any two elements of G, and let $Y \in Aut(G)$ be any automorphism of $(G, +)$. Then, the unique solution of the automorphism equation*

$$Y = -\mathrm{gyr}[b, Xa]X \tag{1.95}$$

for the unknown automorphism $X \in Aut(G)$ is

$$X = -\mathrm{gyr}[b, Ya]Y . \tag{1.96}$$

Proof.   Let X be a solution of (1.95), and let $x \in G$ be given by the equation

$$x = b \boxminus Xa , \tag{1.97}$$

so that, by a right cancellation, (1.71), $b = x + Xa$.

Then we have the following gyrosemidirect product

$$\begin{aligned}
(x, X)(a, I) &= (x + Xa, \mathrm{gyr}[x, Xa]X) \\
&= (x + Xa, \mathrm{gyr}[x + Xa, Xa]X) \\
&= (b, \mathrm{gyr}[b, Xa]X) \\
&= (b, -Y) ,
\end{aligned} \tag{1.98}$$

so that

$$\begin{aligned}
(x, X) &= (b, -Y)(a, I)^{-1} \\
&= (b, -Y)(-a, I) \\
&= (b + Ya, -\mathrm{gyr}[b, Ya]Y) .
\end{aligned} \tag{1.99}$$

Comparing the second entries of the extreme sides of (1.99) we have

$$X = -\mathrm{gyr}[b, Ya]Y . \tag{1.100}$$

Hence, if a solution X of (1.95) exists, then it must be given by (1.96).

Conversely, the automorphism X in (1.100) is, indeed, a solution of (1.95) as we see by substituting X from (1.100) into the right-hand side of (1.95), and employing the nested gyration identity (1.40), p. 13,

$$- \mathrm{gyr}[b, Xa]X = \mathrm{gyr}[b, -\mathrm{gyr}[b, Ya]Ya]\mathrm{gyr}[b, Ya]Y = Y . \tag{1.101}$$

$\square$

1.9   BASIC GYRATION PROPERTIES

We use the notation

$$(\mathrm{gyr}[a, b])^{-1} = \mathrm{gyr}^{-1}[a, b] \tag{1.102}$$

for the inverse gyroautomorphism.

The most important basic gyration properties that we establish in this section are the *gyration even property*

$$\mathrm{gyr}[\ominus a, \ominus b] = \mathrm{gyr}[a, b] \tag{1.103}$$

and the *gyroautomorphism inversion law*

$$\mathrm{gyr}^{-1}[a, b] = \mathrm{gyr}[b, a] \tag{1.104}$$

for any two elements a and b of a gyrogroup $(G, \oplus)$.

Theorem 1.26.   (Gyrosum Inversion, Gyroautomorphism Inversion). *For any two elements a, b of a gyrogroup $(G, +)$ we have the gyrosum inversion law*

$$- (a + b) = \mathrm{gyr}[a, b](-b - a), \tag{1.105}$$

and the gyroautomorphism inversion law

$$\mathrm{gyr}^{-1}[a, b] = \mathrm{gyr}[-b, -a]. \tag{1.106}$$

Proof. Let $Aut_0(G)$ be any gyroautomorphism group of $(G, +)$, and let $G \times Aut_0(G)$ be the gyrosemidirect product of the gyrogroup G and the group $Aut_0(G)$ according to Def. 1.23. Being a group, the product of two elements of the gyrosemidirect product group $G \times Aut_0(G)$ has a unique inverse. This inverse can be calculated in two different ways.

On the one hand, the inverse of the left-hand side of the product

$$(a, I)(b, I) = (a + b, \mathrm{gyr}[a, b]) \tag{1.107}$$

in $G \times Aut_0(G)$ is

$$\begin{aligned} (b, I)^{-1}(a, I)^{-1} &= (-b, I)(-a, I) \\ &= (-b - a, \mathrm{gyr}[-b, -a]). \end{aligned} \tag{1.108}$$

On the other hand, the inverse of the right-hand side of the product (1.107) is, by (1.91),

$$(-\mathrm{gyr}^{-1}[a, b](a + b), \ \mathrm{gyr}^{-1}[a, b]) \tag{1.109}$$

for all $a, b \in G$. Comparing corresponding entries in (1.108) and (1.109) we have

$$- b - a = -\mathrm{gyr}^{-1}[a, b](a + b) \tag{1.110}$$

and

$$\mathrm{gyr}[-b, -a] = \mathrm{gyr}^{-1}[a, b]. \tag{1.111}$$

Eliminating $\mathrm{gyr}^{-1}[a, b]$ between (1.110) and (1.111), we have

$$-b - a = -\mathrm{gyr}[-b, -a](a + b). \tag{1.112}$$

Replacing (a, b) by $(-b, -a)$, (1.112) becomes

$$a + b = -\mathrm{gyr}[a, b](-b - a). \tag{1.113}$$

Identities (1.113) and (1.111) complete the proof.   □

Instructively, the gyrosum inversion law (1.105) is verified here as a by-product along with the gyroautomorphism inversion law (1.106) in Theorem 1.26 in terms of the gyrosemidirect product group. A direct proof of (1.105) is, however, simpler as we saw in Theorem 1.14, p. 12.

Theorem 1.27.   *Let $(G, +)$ be a gyrogroup. Then for all $a, b \in G$*

$$
\begin{aligned}
\mathrm{gyr}^{-1}[a, b] &= \mathrm{gyr}[a, -\mathrm{gyr}[a, b]b] & (1.114)\\
\mathrm{gyr}^{-1}[a, b] &= \mathrm{gyr}[-a, a + b] & (1.115)\\
\mathrm{gyr}^{-1}[a, b] &= \mathrm{gyr}[b, a + b] & (1.116)\\
\mathrm{gyr}[a, b] &= \mathrm{gyr}[b, -b - a] & (1.117)\\
\mathrm{gyr}[a, b] &= \mathrm{gyr}[-a, -b - a] & (1.118)\\
\mathrm{gyr}[a, b] &= \mathrm{gyr}[-(a + b), a]. & (1.119)
\end{aligned}
$$

Proof. Identity (1.114) follows from (1.40).

Identity (1.115) follows from (1.41).

Identity (1.116) results from an application to (1.115) of the left loop property followed by a left cancellation.

Identity (1.117) follows from the gyroautomorphism inversion law (1.106) and from (1.115),

$$\mathrm{gyr}[a, b] = \mathrm{gyr}^{-1}[-b, -a] = \mathrm{gyr}[b, -b - a]. \tag{1.120}$$

Identity (1.118) follows from an application, to the right-hand side of (1.117), of the left loop property followed by a left cancellation.

Identity (1.119) follows by inverting (1.115) by means of the gyroautomorphism inversion law (1.106).   □

Theorem 1.28.   (The Gyration Inversion Law; The Gyration Even Property). *The gyroautomorphisms of any gyrogroup $(G, +)$ obey the gyration inversion law*

$$\mathrm{gyr}^{-1}[a, b] = \mathrm{gyr}[b, a], \tag{1.121}$$

and possess the gyration even property

$$\text{gyr}[-a, -b] = \text{gyr}[a, b],\tag{1.122}$$

satisfying the four mutually equivalent nested gyroautomorphism identities

$$\begin{aligned}
\text{gyr}[b, -\text{gyr}[b, a]a] &= \text{gyr}[a, b] \\
\text{gyr}[b, \text{gyr}[b, -a]a] &= \text{gyr}[a, -b] \\
\text{gyr}[-\text{gyr}[a, b]b, a] &= \text{gyr}[a, b] \\
\text{gyr}[\text{gyr}[a, -b]b, a] &= \text{gyr}[a, -b]
\end{aligned}\tag{1.123}$$

for all $a, b \in G$.

Proof. By the left loop property and (1.116) we have

$$\begin{aligned}
\text{gyr}^{-1}[a + b, b] &= \text{gyr}^{-1}[a, b] \\
&= \text{gyr}[b, a + b]
\end{aligned}\tag{1.124}$$

for all $a, b \in G$. Let us substitute $a = c \boxminus b$ into (1.124), so that by a right cancellation $a + b = c$, obtaining the identity

$$\text{gyr}^{-1}[c, b] = \text{gyr}[b, c]\tag{1.125}$$

for all $c, b \in G$. Renaming c in (1.125) as a, we obtain (1.121), as desired.

Identity (1.122) results from (1.106) and (1.121),

$$\begin{aligned}
\text{gyr}[-a, -b] &= \text{gyr}^{-1}[b, a] \\
&= \text{gyr}[a, b].
\end{aligned}\tag{1.126}$$

Finally, the first identity in (1.123) follows from (1.114) and (1.121).

By means of the gyroautomorphism inversion law (1.121), the third identity in (1.123) is equivalent to the first one.

The second (fourth) identity in (1.123) follows from the first (third) by replacing a by $-a$ (or, alternatively, by replacing b by $-b$), noting that gyrations are even by (1.122). $\qquad\square$

The left gyroassociative law and the left loop property of gyrogroups admit right counterparts, as we see from the following theorem.

Theorem 1.29. *For any three elements a, b, and c of a gyrogroup $(G, \oplus)$ we have*

(i) $\quad (a \oplus b) \oplus c = a \oplus (b \oplus \text{gyr}[b, a]c) \qquad$ *Right Gyroassociative Law,*

(ii) $\quad \text{gyr}[a, b] = \text{gyr}[a, b \oplus a] \qquad\qquad$ *Right Loop Property.*

Proof. The right gyroassociative law follows from the left gyroassociative law and the gyration inversion law (1.121) of gyroautomorphisms,

$$a \oplus (b \oplus \text{gyr}[b, a]c) = (a \oplus b) \oplus \text{gyr}[a, b]\text{gyr}[b, a]c$$
$$= (a \oplus b) \oplus c .$$

(1.127)

The right loop property results from (1.116) and the gyration inversion law (1.121),

$$\text{gyr}[b, a + b] = \text{gyr}^{-1}[a, b]$$
$$= \text{gyr}[b, a] .$$

(1.128)

□

The right cancellation law allows the loop property to be dualized in the following theorem.

Theorem 1.30.   (The Coloop Property - Left and Right). *Let* $(G, \oplus)$ *be a gyrogroup. Then*

$$\text{gyr}[a, b] = \text{gyr}[a \boxminus b, b] \qquad \text{Left Coloop Property,}$$
$$\text{gyr}[a, b] = \text{gyr}[a, b \boxminus a] \qquad \text{Right Coloop Property,}$$

for all $a, b \in G$.

Proof. The proof follows from an application of the left and the right loop property followed by a right cancellation,

$$\text{gyr}[a \boxminus b, b] = \text{gyr}[(a \boxminus b) \oplus b, b] = \text{gyr}[a, b] ,$$
$$\text{gyr}[a, b \boxminus a] = \text{gyr}[a, (b \boxminus a) \oplus a] = \text{gyr}[a, b] .$$

(1.129)

□

A right and a left loop give rise to the identities in the following theorem.

**Theorem 1.31.   *Let* $(G, \oplus)$ *be a gyrogroup. Then*

$$\text{gyr}[a \oplus b, \ominus a] = \text{gyr}[a, b] ,$$
$$\text{gyr}[\ominus a, a \oplus b] = \text{gyr}[b, a] ,$$

(1.130)

for all $a, b \in G$.

Proof. By a right loop, a left cancellation and a left loop we have

$$\text{gyr}[a \oplus b, \ominus a] = \text{gyr}[a \oplus b, \ominus a \oplus (a \oplus b)]$$
$$= \text{gyr}[a \oplus b, b]$$
$$= \text{gyr}[a, b] ,$$

(1.131)

thus verifying the first identity in (1.130). The second identity in (1.130) follows from the first one by gyroautomorphism inversion, (1.121).

□

In general, $-(a+b) \neq -a - b$ in a gyrogroup $(G, +)$. In fact, we have $-(a+b) = -a - b$ for all $a, b \in G$ if and only if the gyrogroup $(G, +)$ is gyrocommutative, as we see from Theorem 2.2, p. 33, of Chap. 2 on gyrocommutative gyrogroups. In this sense, the gyrogroup cooperation $\boxplus$ conducts itself more properly than its associated gyrogroup operation, as we see from the following theorem.

Theorem 1.32.   (The Cogyroautomorphic Inverse Property). *Any gyrogroup $(G, +)$ possesses the cogyroautomorphic inverse property,*

$$- (a \boxplus b) = (-b) \boxplus (-a) \tag{1.132}$$

for any $a, b \in G$.

Proof. We verify (1.132) in the following chain of equations, which are numbered for subsequent derivation:

$$
\begin{aligned}
a \boxplus b &\overset{(1)}{=\!=\!=} a + \mathrm{gyr}[a, -b]b \\
&\overset{(2)}{=\!=\!=} -\mathrm{gyr}[a, \mathrm{gyr}[a, -b]b]\{-\mathrm{gyr}[a, -b]b - a\} \\
&\overset{(3)}{=\!=\!=} \mathrm{gyr}[a, \mathrm{gyr}[a, -b]b]\{-(-\mathrm{gyr}[a, -b]b - a)\} \\
&\overset{(4)}{=\!=\!=} \mathrm{gyr}[a, -\mathrm{gyr}[a, -b](-b)]\{-(-\mathrm{gyr}[a, -b]b - a)\} \\
&\overset{(5)}{=\!=\!=} \mathrm{gyr}^{-1}[a, -b]\{-(-\mathrm{gyr}[a, -b]b - a)\} \\
&\overset{(6)}{=\!=\!=} -(-b - \mathrm{gyr}^{-1}[a, -b]a) \\
&\overset{(7)}{=\!=\!=} -\{-b - \mathrm{gyr}[b, -a]a\} \\
&\overset{(8)}{=\!=\!=} -\{(-b) \boxplus (-a)\} .
\end{aligned}
\tag{1.133}
$$

Inverting both extreme sides of (1.133) we obtain the desired identity (1.132).

Derivation of the numbered equalities in (1.133) follows.

(1) Follows from Def. 1.9, p. 7, of the gyrogroup cooperation $\boxplus$.

(2) Follows from (1) by the gyrosum inversion, (1.105).

(3) Follows from (2) by Theorem 1.13(12) applied to the term $\{\ldots\}$ in (2).

(4) Follows from (3) by Theorem 1.13(12) applied to b, that is, $\mathrm{gyr}[a, -b]b = -\mathrm{gyr}[a, -b](-b)$.

(5) Follows from (4) by Identity (1.114) of Theorem 1.27.

(6) Follows from (5) by distributing the gyroautomorphism $\text{gyr}^{-1}[a, -b]$ over each of the two terms in $\{\ldots\}$.

(7) Follows from (6) by the gyroautomorphism inversion law (1.106).

(8) Follows from (7) by Def. 1.9, p. 7, of the gyrogroup cooperation $\boxplus$. $\square$

Theorem 1.33.   *Let* $(G, \oplus)$ *be a gyrogroup. Then*

$$a\oplus\{(\ominus a\oplus b)\oplus a\} = b \boxplus a \tag{1.134}$$

for all $a, b \in G$.

Proof. The proof rests on the following chain of equations, which are numbered for subsequent explanation:

$$
\begin{aligned}
a\oplus\{(\ominus a\oplus b)\oplus a\} &\overset{(1)}{=\!=} \{a\oplus(\ominus a\oplus b)\}\oplus\text{gyr}[a, \ominus a\oplus b]a \\
&\overset{(2)}{=\!=} b\oplus\text{gyr}[b, \ominus a\oplus b]a \\
&\overset{(3)}{=\!=} b\oplus\text{gyr}[b, \ominus a]a \\
&\overset{(4)}{=\!=} b \boxplus a \, .
\end{aligned}
\tag{1.135}
$$

The derivation of the equalities in (1.135) follows.

(1) Follows from the left gyroassociative law.

(2) Follows from (1) by a left cancellation, and by a left loop followed by a left cancellation.

(3) Follows from (2) by a right loop, that is, an application of the right loop property to (3) gives (2).

(4) Follows from (3) by Def. 1.9, p. 7, of the gyrogroup cooperation $\boxplus$. $\square$

1.10   AN ADVANCED GYROGROUP EQUATION

As an example, and for later convenience, we present in Theorem 1.34 below an advanced gyrogroup equation and its unique solution. The equation is advanced in the sense that its unknown appears in the equation both directly, and indirectly in the argument of a gyration.

Theorem 1.34.   *Let*

$$c = \text{gyr}[b, -x]x \tag{1.136}$$

be an equation for the unknown x *in a gyrogroup* $(G, +)$. *The unique solution of* (1.136) *is*

$$x = -(-b - (c \boxminus b)) \, . \tag{1.137}$$

Proof. If a solution x to the gyrogroup equation (1.136) exists then, by (1.136) and by the second identity in (1.37), p. 13,

$$c = \mathrm{gyr}[b, -x]x = -(b - x) + b. \tag{1.138}$$

Applying a right cancellation to (1.138) we obtain

$$-(b - x) = c \boxminus b, \tag{1.139}$$

or, equivalently, by gyro-sign inversion,

$$b - x = -(c \boxminus b), \tag{1.140}$$

so that, by a left cancellation

$$-x = -b - (c \boxminus b), \tag{1.141}$$

implying, by gyro-sign inversion,

$$x = -(-b - (c \boxminus b)). \tag{1.142}$$

Hence, if a solution x to (1.136) exists, it must be given by (1.142).

Conversely, x given by (1.142) is a solution. Indeed, the substitution of x of (1.142) into the right-hand side of (1.138) results in c, as we see in the following chain of equations, which are numbered for subsequent derivation:

$$
\begin{aligned}
\mathrm{gyr}[b, -x]x \overset{(1)}{=\!=\!=}\ & -\mathrm{gyr}[b, -b - (c \boxminus b)]\{-b - (c \boxminus b)\} \\[4pt]
\overset{(2)}{=\!=\!=}\ & -\{b + (-b - (c \boxminus b))\} + b \\[4pt]
\overset{(3)}{=\!=\!=}\ & -\mathrm{gyr}[-\{b + (-b - (c \boxminus b))\}, b](-b + \{b + (-b - (c \boxminus b))\}) \\[4pt]
\overset{(4)}{=\!=\!=}\ & -\mathrm{gyr}[b + (-b - (c \boxminus b)), -b]\{-b - (c \boxminus b)\} \\[4pt]
\overset{(5)}{=\!=\!=}\ & -\mathrm{gyr}[b, -b - (c \boxminus b)]\{-b - (c \boxminus b)\} \\[4pt]
\overset{(6)}{=\!=\!=}\ & -\mathrm{gyr}[-(c \boxminus b), -b]\{-b - (c \boxminus b)\} \\[4pt]
\overset{(7)}{=\!=\!=}\ & -\mathrm{gyr}[c \boxminus b, b]\{-b - (c \boxminus b)\} \\[4pt]
\overset{(8)}{=\!=\!=}\ & (c \boxminus b) + b \\[4pt]
\overset{(9)}{=\!=\!=}\ & c.
\end{aligned}
\tag{1.143}
$$

Derivation of the numbered equalities in (1.143) follows.

(1) Follows from the substitution of x from (1.142), and from Theorem 1.13(12).

(2) Follows from (1) by the first identity in (1.37), p. 13.

(3) Follows from (2) by the gyrosum inversion law (1.34), p. 12.

(4) Follows from (3) by the gyration even property and by a left cancellation.

(5) Follows from (4) by the first identity in (1.130).

(6) Follows from (5) by the second identity in (1.130).

(7) Follows from (6) by the gyration even property.

(8) Follows from (7) by the gyrosum inversion law (1.34), p. 12.

(9) Follows from (8) by a right cancellation.

$\square$

Corollary 1.35.   Let $(G, +)$ be a gyrogroup. The map

$$a \mapsto c = \text{gyr}[b, -a]a \tag{1.144}$$

of G into itself is bijective so that when a runs over all the elements of G its image, c, runs over all the elements of G as well for any given element $b \in G$.

Proof. It follows immediately from Theorem 1.34 that, for any given $b \in G$, the map (1.144) is bijective and, hence, the result of the Corollary. $\square$

1.11 EXERCISES

(1) Being automorphisms by an axiom, the gyrations of a gyrogroup $(G, \oplus)$ preserve the gyrogroup operation $\oplus$, that is

$$\text{gyr}[a, b](u \oplus v) = \text{gyr}[a, b]u \oplus \text{gyr}[a, b]v \tag{1.145}$$

for all $a, b, u, v \in G$.

Prove that, like the group operation $\oplus$, also the gyrogroup cooperation $\boxplus$ of a gyrogroup $(G, \oplus)$ is preserved by gyrations, that is,

$$\text{gyr}[a, b](u \boxplus v) = \text{gyr}[a, b]u \boxplus \text{gyr}[a, b]v \tag{1.146}$$

for all $a, b, u, v \in G$.

Hint: Apply Theorem 1.21, p. 20, to $u \boxplus v$ in Def. 1.9, p. 7.

CHAPTER 2

Gyrocommutative Gyrogroups

Some gyrocommutative gyrogroups give rise to gyrovector spaces, which are the framework for analytic hyperbolic geometry just as some commutative groups give rise to vector spaces, which are the framework for analytic Euclidean geometry. To pave the way for gyrovector spaces we, therefore, study gyrocommutative gyrogroups in this chapter. In gyrocommutative gyrogroups the gyrogroup operation is gyrocommutative, by Def. 1.8, p. 6, while the gyrogroup cooperation is commutative, by Theorem 2.3 below.

2.1 GYROCOMMUTATIVE GYROGROUPS

Definition 2.1. (Gyroautomorphic Inverse Property). *A gyrogroup* $(G, +)$ *possesses the gyroautomorphic inverse property if for all* $a, b \in G$,

$$- (a + b) = -a - b. \qquad (2.1)$$

Theorem 2.2. (The Gyroautomorphic Inverse Property). *A gyrogroup is gyrocommutative if and only if it possesses the gyroautomorphic inverse property.*

Proof. Let $(G, +)$ be a gyrogroup possessing the gyroautomorphic inverse property. Then the gyrosum inversion law (1.34), p. 12, specializes, by means of Theorem 1.13(12), p. 11, to the gyrocommutative law $(G6)$ in Def. 1.8, p. 6,

$$
\begin{aligned}
a + b &= -\text{gyr}[a, b](-b - a) \\
&= \text{gyr}[a, b]\{-(-b - a)\} \\
&= \text{gyr}[a, b](b + a)
\end{aligned}
\qquad (2.2)
$$

for all $a, b \in G$.

Conversely, if the gyrocommutative law is valid then by Theorem 1.13(12) and the gyrosum inversion law, (1.34), p. 12, we have

$$\text{gyr}[a, b]\{-(-b - a)\} = -\text{gyr}[a, b](-b - a) = a + b = \text{gyr}[a, b](b + a), \qquad (2.3)$$

so that by eliminating the gyroautomorphism $\text{gyr}[a, b]$ on both extreme sides of (2.3) and inverting the gyro-sign we recover the gyroautomorphic inverse property,

$$- (b + a) = -b - a \qquad (2.4)$$

for all $a, b \in G$. $\qquad \square$

Theorem 2.3.   *The gyrogroup cooperation $\boxplus$ of a gyrogroup $(G, +)$ is commutative if and only if the gyrogroup $(G, +)$ is gyrocommutative.*

Proof. For any $a, b \in G$ we have, by equality (7) of the chain of equations (1.133) in the proof of Theorem 1.32, p. 29,

$$a \boxplus b = -(-b - \text{gyr}[b, -a]a) . \tag{2.5}$$

But by definition,

$$b \boxplus a = b + \text{gyr}[b, -a]a . \tag{2.6}$$

Hence,

$$a \boxplus b = b \boxplus a \tag{2.7}$$

for all $a, b \in G$ if and only if

$$-(-b - c) = b + c \tag{2.8}$$

for all $a, b \in G$, where

$$c = \text{gyr}[b, -a]a , \tag{2.9}$$

as we see from (2.5) and (2.6). But the self-map of G that takes a to c in (2.9),

$$a \mapsto \text{gyr}[b, -a]a = c \tag{2.10}$$

for any given $b \in G$ is bijective, by Corollary 1.35, p. 32. Hence, the commutative relation (2.7) for $\boxplus$ holds for all $a, b \in G$ if and only if (2.8) holds for all $b, c \in G$. The latter, in turn, is the gyroautomorphic inverse property that, by Theorem 2.2, is equivalent to the gyrocommutativity of the gyrogroup $(G, +)$. Hence, (2.7) holds for all $a, b \in G$ if and only if the gyrogroup $(G, +)$ is gyrocommutative. $\qquad \square$

Theorem 2.4.   *Let $(G, +)$ be a gyrocommutative gyrogroup. Then*

$$\text{gyr}[a, b]\text{gyr}[b + a, c] = \text{gyr}[a, b + c]\text{gyr}[b, c] \tag{2.11}$$

for all $a, b, c \in G$.

Proof. Using the notation $g_{a,b} = \text{gyr}[a, b]$ whenever convenient, we have by Theorem 1.21, p. 20, by the gyrocommutative law, and by Identity (1.38), p. 13,

$$\begin{aligned}
\text{gyr}[a, b]\text{gyr}[b + a, c] &= \text{gyr}[g_{a,b}(b + a), g_{a,b}c]\text{gyr}[a, b] \\
&= \text{gyr}[a + b, \text{gyr}[a, b]c]\text{gyr}[a, b] \\
&= \text{gyr}[a, b + c]\text{gyr}[b, c] .
\end{aligned} \tag{2.12}$$

$\qquad \square$

Theorem 2.5.   *Let $a, b, c \in G$ be any three elements of a gyrocommutative gyrogroup $(G, +)$, and let $d \in G$ be determined by the "gyroparallelogram condition"*

$$d = (b \boxplus c) - a .   \tag{2.13}$$

Then, the elements $a, b, c,$ and d satisfy the telescopic gyration identity

$$\text{gyr}[a, -b]\text{gyr}[b, -c]\text{gyr}[c, -d] = \text{gyr}[a, -d]   \tag{2.14}$$

for all $a, b, c \in G$.

Proof. By Identity (2.11), along with an application of the right and the left loop property, we have

$$\text{gyr}[a', b' + a']\text{gyr}[b' + a', c'] = \text{gyr}[a', b' + c']\text{gyr}[b' + c', c'] .   \tag{2.15}$$

Let

$$a = -c'$$
$$c = -a'   \tag{2.16}$$
$$b = b' + a' ,$$

so that, by the third equation in (2.16), and by (2.13) we have,

$$\begin{aligned} b' + c' &= (b \boxminus a') + c' \\ &= (b \boxplus c) - a \\ &= d . \end{aligned}   \tag{2.17}$$

Then (2.15), expressed in terms of a, b, c, d in (2.16)–(2.17), takes the form

$$\text{gyr}[-c, b]\text{gyr}[b, -a] = \text{gyr}[-c, d]\text{gyr}[d, -a] .   \tag{2.18}$$

Inverting both sides of (2.18) by means of the gyration inversion law (1.121), p. 26, and the gyration even property (1.122), p. 27, we obtain the identity

$$\text{gyr}[a, -b]\text{gyr}[b, -c] = \text{gyr}[a, -d]\text{gyr}[d, -c] ,   \tag{2.19}$$

from which the telescopic gyration identity (2.14) follows immediately, by a gyration inversion and the gyration even property.   □

Theorem 2.6.   *The gyroparallelogram condition (2.13),*

$$d = (b \boxplus c) - a ,   \tag{2.20}$$

is equivalent to the identity

$$-c + d = \text{gyr}[c, -b](b - a) .   \tag{2.21}$$

Proof. In a gyrocommutative gyrogroup $(G, +)$ the gyroparallelogram condition (2.13) implies the following chain of equations, which are numbered for subsequent derivation:

$$
\begin{aligned}
d &\overset{(1)}{=\!=\!=} (b \boxplus c) - a \\
&\overset{(2)}{=\!=\!=} (c \boxplus b) - \mathrm{gyr}[b, -c]\mathrm{gyr}[c, -b]a \\
&\overset{(3)}{=\!=\!=} (c \boxplus b) - \mathrm{gyr}[c, \mathrm{gyr}[c, -b]b]\mathrm{gyr}[c, -b]a \\
&\overset{(4)}{=\!=\!=} (c + \mathrm{gyr}[c, -b]b) - \mathrm{gyr}[c, \mathrm{gyr}[c, -b]b]\mathrm{gyr}[c, -b]a \\
&\overset{(5)}{=\!=\!=} c + (\mathrm{gyr}[c, -b]b - \mathrm{gyr}[c, -b]a) \\
&\overset{(6)}{=\!=\!=} c + \mathrm{gyr}[c, -b](b - a)
\end{aligned}
\tag{2.22}
$$

for all $a, b, c \in G$. Derivation of the numbered equalities in (2.22) follows.

(1) This equation is the gyroparallelogram condition (2.13).

(2) Follows from (1) (i) since the gyrogroup cooperation $\boxplus$ is commutative, by Theorem 2.3, p. 34, and (ii) since, by gyration inversion along with the gyration even property in (1.121)–(1.122), p. 27, the gyration product applied to a in (2) is trivial.

(3) Follows from (2) by the second nested gyration identity in (1.123), p. 27.

(4) Follows from (3) by Def. 1.9, p. 7, of the gyrogroup cooperation $\boxplus$.

(5) Follows from (4) by the left gyroassociative law. Indeed, an application of the left gyroassociative law to (5) results in (4).

(6) Follows from (5) since gyroautomorphisms respect their gyrogroup operation.

Finally, (2.21) follows from (2.22) by a left cancellation, moving c from the extreme right-hand side of (2.22) to its extreme left-hand side. □

Theorem 2.7.   *Let $(G, +)$ be a gyrocommutative gyrogroup. Then*

$$
\mathrm{gyr}[a, b]\{b + (a + c)\} = (a + b) + c \tag{2.23}
$$

for all $a, b, c \in G$.

Proof. By the left gyroassociative law and by the gyrocommutative law we have the chain of equations

$$
\begin{aligned}
b + (a + c) &= (b + a) + \mathrm{gyr}[b, a]c \\
&= \mathrm{gyr}[b, a](a + b) + \mathrm{gyr}[b, a]c \\
&= \mathrm{gyr}[b, a]\{(a + b) + c\},
\end{aligned}
\tag{2.24}
$$

from which (2.23) is derived by the gyroautomorphism inversion law (1.106), p. 25. □

The special case of Theorem 2.7 corresponding to $c = -a$ gives rise to a new cancellation law in gyrocommutative gyrogroups, called the *left-right cancellation law*.

Theorem 2.8.   (The Left-Right Cancellation Law). *Let $(G, +)$ be a gyrocommutative gyrogroup. Then*

$$(a + b) - a = \text{gyr}[a, b]b \tag{2.25}$$

for all $a, b, c \in G$.

Proof. Identity (2.25) follows from (1.37), p. 13, and the gyroautomorphic inverse property (2.1), p. 33. Alternatively, Identity (2.25) is equivalent to the special case of (2.24) when $c = -a$.   □

The left-right cancellation law (2.25) is not a complete cancellation since the echo of the "canceled" a remains in the argument of the involved gyroautomorphism.

Theorem 2.9. *Let $(G, +)$ be a gyrocommutative gyrogroup. Then*

$$a + \{(-a + b) + a\} = a \boxplus b \tag{2.26}$$

for all $a, b \in G$.

Proof. By Theorem 1.33, p. 30, and Theorem 2.3,

$$a + \{(-a + b) + a\} = b \boxplus a = a \boxplus b. \tag{2.27}$$

□

Theorem 2.10. *Let $(G, +)$ be a gyrocommutative gyrogroup. Then*

$$a \boxplus (a + b) = a + (b + a) \tag{2.28}$$

for all $a, b \in G$.

Proof. By a left cancellation and Theorem 2.9 we have

$$a + (b + a) = a + (\{-a + (a + b)\} + a) = a \boxplus (a + b). \tag{2.29}$$

□

Theorem 2.11.   (The Gyrotranslation Theorem, II). *Let $(G, +)$ be a gyrocommutative gyrogroup. For all $a, b, c \in G$,*

$$\begin{aligned} -(a + b) + (a + c) &= \text{gyr}[a, b](-b + c) \\ (a + b) - (a + c) &= \text{gyr}[a, b](b - c). \end{aligned} \tag{2.30}$$

Proof. The first identity in (2.30) follows from the Gyrotranslation Theorem 1.19, p. 16, with a replaced by $-a$. Hence, it is valid in nongyrocommutative gyrogroups as well. The second identity in (2.30) follows from the first by the gyroautomorphic inverse property of gyrocommutative gyrogroups, Theorem 2.2, p. 33. Hence, it is valid in gyrocommutative gyrogroups.   □

Remark 2.12.   *We should note that, unexpectedly, the gyrations in (2.30) are independent of c. This observation proves useful in the demonstration that gyroangles in gyrovector spaces are invariant under left gyrotranslations, Theorem 4.3, p. 112. The classical counterparts of identities (2.30) in commutative groups are immediate. However, they play important role in the translation of vectors in Euclidean geometry. In full analogy, we will find in Chaps. 3 and 4 that identities (2.30) of the Gyrotranslation Theorem 2.11 in gyrocommutative gyrogroups play important role in the hyperbolic translation of hyperbolic vectors in hyperbolic geometry, where they are called gyrovectors. Indeed, classically, vectors are invariant under Euclidean translations. In contrast, the identities in (2.30) indicate that gyrovectors are not invariant under hyperbolic translations. Rather, under hyperbolic translations they are distorted by gyrations. However, taking gyrations and their structure into account, we will construct in Sec. 3.15 special hyperbolic translations, called gyrovector gyrotranslations, that keep gyrovectors invariant. The consequences of Theorem 2.11 are, thus, far reaching.*

The following theorem gives an elegant gyration identity in which the product of three telescopic gyrations is equivalent to a single gyration. The identity of this theorem is useful in the proof of Theorem 2.15 which, in turn, leads to the gyroparallelogram addition law of gyrovectors in Figs. 3.9, p. 97, and 3.10, p. 98, as we will see in Chap. 3.

Theorem 2.13.   *Let $a, b, c \in G$ be any three elements of a gyrocommutative gyrogroup $(G, +)$. Then,*

$$\mathrm{gyr}[-a + b, a - c] = \mathrm{gyr}[a, -b]\mathrm{gyr}[b, -c]\mathrm{gyr}[c, -a]\,. \tag{2.31}$$

Proof. By Theorem 1.21, p. 20, and by the gyrocommutative law we have

$$\begin{aligned} \mathrm{gyr}[a, b]\mathrm{gyr}[b + a, c] &= \mathrm{gyr}[\mathrm{gyr}[a, b](b + a), \mathrm{gyr}[a, b]c]\mathrm{gyr}[a, b] \\ &= \mathrm{gyr}[a + b, \mathrm{gyr}[a, b]c]\mathrm{gyr}[a, b]\,. \end{aligned} \tag{2.32}$$

Hence, Identity (1.38) in Theorem 1.16, p. 13, can be written as

$$\mathrm{gyr}[a, b + c]\mathrm{gyr}[b, c] = \mathrm{gyr}[a, b]\mathrm{gyr}[b + a, c]\,. \tag{2.33}$$

By gyroautomorphism inversion, the latter can be written as

$$\mathrm{gyr}[a, b + c] = \mathrm{gyr}[a, b]\mathrm{gyr}[b + a, c]\mathrm{gyr}[c, b]\,. \tag{2.34}$$

Using the notation $b + a = d$, which implies $a = -b + d$, Identity (2.34) becomes, by means of Theorem 1.31, p. 28,

$$\text{gyr}[-b + d, b + c] = \text{gyr}[-b + d, b]\text{gyr}[d, c]\text{gyr}[c, b]$$
$$= \text{gyr}[-b, d]\text{gyr}[d, c]\text{gyr}[c, b]\,. \tag{2.35}$$

Renaming the elements $b, c, d \in G$, $(b, c, d) \to (-a, c, -b)$, (2.35) becomes

$$\text{gyr}[a - b, -a + c] = \text{gyr}[a, -b]\text{gyr}[-b, c]\text{gyr}[c, -a]\,. \tag{2.36}$$

By means of the gyroautomorphic inverse property, Theorem 2.2, p. 33, and the gyration even property (1.122) in Theorem 1.28, p. 26, Identity (2.36) can be written, finally, in the desired form (2.31). □

The special case of Theorem 2.13 when $c = -b$ is interesting, giving rise to the following theorem.

Theorem 2.14.   *Let $(G, +)$ be a gyrocommutative gyrogroup. Then*

$$\text{gyr}[a, -b] = \text{gyr}[-a + b, a + b]\text{gyr}[a, b]\,. \tag{2.37}$$

Proof. Owing to the gyration inversion law in Theorem 1.28, p. 26, Identity (2.31) can be written as

$$\text{gyr}[-a + b, a - c]\text{gyr}[-a, c] = \text{gyr}[a, -b]\text{gyr}[b, -c]\,, \tag{2.38}$$

from which the result (2.37) of the theorem follows in the special case when $c = -b$ by applying the gyration even property. □

Identity (2.37) is interesting since it relates the gyration $\text{gyr}[a, -b]$ to the gyration $\text{gyr}[a, b]$. Furthermore, it gives rise, by gyration inversion and the gyration even property, to the following elegant gyration identity,

$$\text{gyr}[-a, b]\text{gyr}[b, a] = \text{gyr}[-a + b, a + b]\,, \tag{2.39}$$

in which the product of two gyrations is equivalent to a single gyration.

As an application of Theorem 2.13, and for later reference, we present the following Theorem 2.15. This theorem, the proof of which involves a long chain of gyrocommutative gyrogroup identities, will prove crucially important for the introduction of hyperbolic vectors, called gyrovectors, into hyperbolic geometry in Chap. 3.

Theorem 2.15.   *Let $(G, +)$ be a gyrocommutative gyrogroup. Then*

$$(x + a) \boxplus (x + b) = x + \{(a \boxplus b) + x\} \tag{2.40}$$

for all $a, b, x \in G$.

Proof. The proof is given by the following chain of equations, which are numbered for subsequent derivation:

$$
\begin{aligned}
(x+a) \boxplus (x+b) &\overset{(1)}{=\!=\!=} (x+a) + \text{gyr}[x+a, -(x+b)](x+b) \\
&\overset{(2)}{=\!=\!=} (x+a) + \text{gyr}[x+a, -x-b](x+b) \\
&\overset{(3)}{=\!=\!=} x + \{a + \text{gyr}[a,x]\text{gyr}[x+a, -x-b](x+b)\} \\
&\overset{(4)}{=\!=\!=} x + \{a + \text{gyr}[a,x]\text{gyr}[-x,-a]\text{gyr}[a,-b]\text{gyr}[b,x](x+b)\} \\
&\overset{(5)}{=\!=\!=} x + \{a + \text{gyr}[a,x]\text{gyr}[x,a]\text{gyr}[a,-b](b+x)\} \\
&\overset{(6)}{=\!=\!=} x + \{a + \text{gyr}[a,-b](b+x)\} \\
&\overset{(7)}{=\!=\!=} x + \{a + (\text{gyr}[a,-b]b + \text{gyr}[a,-b]x)\} \\
&\overset{(8)}{=\!=\!=} x + \{(a + \text{gyr}[a,-b]b) + \text{gyr}[a, \text{gyr}[a,-b]b]\text{gyr}[a,-b]x\} \\
&\overset{(9)}{=\!=\!=} x + \{(a + \text{gyr}[a,-b]b) + \text{gyr}[b,-a]\text{gyr}[a,-b]x\} \\
&\overset{(10)}{=\!=\!=} x + \{(a + \text{gyr}[a,-b]b) + x\} \\
&\overset{(11)}{=\!=\!=} x + \{(a \boxplus b) + x\}.
\end{aligned}
\tag{2.41}
$$

Derivation of the numbered equalities in (2.41) follows.

(1) Follows from Def. 1.9, p. 7, of the gyrogroup cooperation $\boxplus$.

(2) Follows from (1) by the gyroautomorphic inverse property, Theorem 2.2, p. 33.

(3) Follows from (2) by the right gyroassociative law.

(4) Follows from (3) by Identity (2.31) of Theorem 2.13, thus providing an elegant example for an application of that theorem.

(5) Follows from (4) by the gyration even property, and by the gyrocommutative law.

(6) Follows from (5) by the gyration inversion law (1.121), p. 26.

(7) Follows from (6) by expanding the gyration application term by term.

(8) Follows from (7) by the left gyroassociative law.

(9) Follows from (8) by the second identity in (1.123), p. 27.

(10) Follows from (9) by the gyration even property and the gyration inversion law in Theorem 1.28, p. 26, implying $\text{gyr}[b,-a]\text{gyr}[a,-b] = I$.

(11) Follows from (10) by Def. 1.9, p. 7, of the gyrogroup cooperation $\boxplus$. $\square$

   Identity (2.40) of Theorem 2.15 will prove useful in the study of the gyroparallelogram law in Theorem 3.41, p. 93.

2.2   MÖBIUS GYROGROUPS

As suggested in Sec. 1.3, Möbius addition $\oplus_M$ in the ball $\mathbb{V}_s$ of any real inner product space gives rise to a gyrocommutative gyrogroup $(\mathbb{V}_s, \oplus_M)$ called a Möbius gyrogroup. It is assumed that the reader is familiar with real inner product spaces, the formal definition of which will be presented in Def. 3.1 in Chap. 3, p. 55. The definition of Möbius addition, $\oplus_M$, in the ball follows.

Definition 2.16.   (Möbius Addition in the Ball). *Let* $\mathbb{V} = (\mathbb{V}, +, \cdot)$ *be a real inner product space with a binary operation* $+$ *and a positive definite inner product* $\cdot$ *([28, p. 21]; following [24], also known as Euclidean space) and let* $\mathbb{V}_s$ *be the s-ball of* $\mathbb{V}$,

$$\mathbb{V}_s = \{\mathbf{v} \in \mathbb{V} : \|\mathbf{v}\| < s\} \tag{2.42}$$

for any fixed $s > 0$. *Möbius addition* $\oplus_M$ *is a binary operation in* $\mathbb{V}_s$ *given by the equation*

$$\mathbf{u}\oplus_M\mathbf{v} = \frac{(1 + \frac{2}{s^2}\mathbf{u}\cdot\mathbf{v} + \frac{1}{s^2}\|\mathbf{v}\|^2)\mathbf{u} + (1 - \frac{1}{s^2}\|\mathbf{u}\|^2)\mathbf{v}}{1 + \frac{2}{s^2}\mathbf{u}\cdot\mathbf{v} + \frac{1}{s^4}\|\mathbf{u}\|^2\|\mathbf{v}\|^2} \tag{2.43}$$

where $\cdot$ *and* $\|\cdot\|$ *are the inner product and norm that the ball* $\mathbb{V}_s$ *inherits from its space* $\mathbb{V}$.

In the limit of large s, $s \to \infty$, the ball $\mathbb{V}_s$ in Def. 2.16 expands to the whole of its space $\mathbb{V}$, and Möbius addition in $\mathbb{V}_s$ reduces to vector addition in $\mathbb{V}$.

Möbius addition satisfies the Möbius gamma identity

$$\gamma_{\mathbf{u}\oplus_M\mathbf{v}} = \gamma_\mathbf{u}\,\gamma_\mathbf{v}\,\sqrt{1 + \frac{2}{s^2}\mathbf{u}\cdot\mathbf{v} + \frac{1}{s^4}\|\mathbf{u}\|^2\|\mathbf{v}\|^2} \tag{2.44}$$

for all $\mathbf{u}, \mathbf{v} \in \mathbb{V}_s$, where $\gamma_\mathbf{v}$ is the gamma factor

$$\gamma_\mathbf{v} = \frac{1}{\sqrt{1 - \dfrac{\|\mathbf{v}\|^2}{s^2}}}, \tag{2.45}$$

which is a real number for any $\mathbf{v}$ in the s-ball $\mathbb{V}_s$.

The gamma factor appears also in Einstein addition, and it is known in special relativity theory as the Lorentz gamma factor. A frequently used identity that follows from (2.45) is

$$\frac{\mathbf{v}^2}{s^2} = \frac{\gamma_\mathbf{v}^2 - 1}{\gamma_\mathbf{v}^2}, \tag{2.46}$$

where we use the notation $\mathbf{v}^2 = \mathbf{v}\cdot\mathbf{v} = \|\mathbf{v}\|^2$.

Möbius gyrations

$$\text{gyr}[\mathbf{u}, \mathbf{v}] : \mathbb{V}_s \to \mathbb{V}_s \tag{2.47}$$

are automorphisms of the Möbius gyrogroup $(\mathbb{V}_s, \oplus_{\mathrm{M}})$,

$$\mathrm{gyr}[\mathbf{u}, \mathbf{v}] \in Aut(\mathbb{V}_s, \oplus_{\mathrm{M}}), \tag{2.48}$$

given by the gyrator identity, Theorem 1.13(10), p. 11,

$$\mathrm{gyr}[\mathbf{u}, \mathbf{v}]\mathbf{w} = \ominus_{\mathrm{M}}(\mathbf{u}\oplus_{\mathrm{M}}\mathbf{v})\oplus_{\mathrm{M}}\{\mathbf{u}\oplus_{\mathrm{M}}(\mathbf{v}\oplus_{\mathrm{M}}\mathbf{w})\}, \tag{2.49}$$

and are further specified in (1.27)–(1.29), p. 9.

Gyrations preserve the inner product that the ball $\mathbb{V}_s$ inherits from its real inner product space $\mathbb{V}$,

$$\mathrm{gyr}[\mathbf{u}, \mathbf{v}]\mathbf{a}\cdot\mathrm{gyr}[\mathbf{u}, \mathbf{v}]\mathbf{b} = \mathbf{a}\cdot\mathbf{b} \tag{2.50}$$

for all $\mathbf{a}, \mathbf{b}, \mathbf{u}, \mathbf{v}, \mathbf{w} \in \mathbb{V}_s$. Hence, in particular, Möbius gyrations preserve the norm that the ball $\mathbb{V}_s$ inherits from its real inner product space $\mathbb{V}$ so that, by the gyrocommutative law,

$$\|\mathbf{u}\oplus_{\mathrm{M}}\mathbf{v}\| = \|\mathrm{gyr}[\mathbf{u}, \mathbf{v}](\mathbf{v}\oplus_{\mathrm{M}}\mathbf{u})\| = \|\mathbf{v}\oplus_{\mathrm{M}}\mathbf{u}\|. \tag{2.51}$$

Furthermore, any gyration of a Möbius gyrogroup $(\mathbb{V}_s, \oplus_{\mathrm{M}})$ can be extended to an invertible linear map of the carrier space $\mathbb{V}$ of the ball $\mathbb{V}_s$, given by (1.27), p. 9.

Möbius gyrogroup cooperation (1.15), p. 3, is given by Möbius coaddition, $\boxplus_{\mathrm{M}}$,

$$\mathbf{u}\boxplus_{\mathrm{M}}\mathbf{v} = \frac{\gamma_{\mathbf{u}}^2\mathbf{u} + \gamma_{\mathbf{v}}^2\mathbf{v}}{\gamma_{\mathbf{u}}^2 + \gamma_{\mathbf{v}}^2 - 1} = \frac{(1 - \|\mathbf{v}\|^2)\mathbf{u} + (1 - \|\mathbf{u}\|^2)\mathbf{v}}{1 - \|\mathbf{u}\|^2\|\mathbf{v}\|^2}, \tag{2.52}$$

satisfying the gamma identity

$$\gamma_{\mathbf{u}\boxplus_{\mathrm{M}}\mathbf{v}} = \frac{\gamma_{\mathbf{u}}^2 + \gamma_{\mathbf{v}}^2 - 1}{\sqrt{1 + 2\gamma_{\mathbf{u}}^2\gamma_{\mathbf{v}}^2(1 - \frac{\mathbf{u}\cdot\mathbf{v}}{s^2}) - (\gamma_{\mathbf{u}}^2 + \gamma_{\mathbf{v}}^2)}}. \tag{2.53}$$

Möbius coaddition is commutative, as expected from Theorem 2.3, p. 34.

When the nonzero vectors $\mathbf{u}$ and $\mathbf{v}$ in the ball $\mathbb{V}_s$ of $\mathbb{V}$ are parallel in $\mathbb{V}$, $\mathbf{u}\|\mathbf{v}$, that is, $\mathbf{u} = \lambda\mathbf{v}$ for some $0 \neq \lambda \in \mathbb{R}$, Möbius addition reduces to the binary operation

$$\mathbf{u}\boxplus_{\mathrm{M}}\mathbf{v} = \mathbf{u}\oplus_{\mathrm{M}}\mathbf{v} = \frac{\mathbf{u} + \mathbf{v}}{1 + \frac{1}{s^2}\|\mathbf{u}\|\|\mathbf{v}\|}, \qquad \mathbf{u}\|\mathbf{v}, \tag{2.54}$$

which is both commutative and associative. We should note that in the special case when $\mathbb{V}$ is a one-dimensional vector space, Möbius addition in (2.54) is a binary operation between real numbers. Accordingly, by (2.54),

$$\|\mathbf{u}\|\boxplus_{\mathrm{M}}\|\mathbf{v}\| = \|\mathbf{u}\|\oplus_{\mathrm{M}}\|\mathbf{v}\| = \frac{\|\mathbf{u}\| + \|\mathbf{v}\|}{1 + \frac{1}{s^2}\|\mathbf{u}\|\|\mathbf{v}\|} \tag{2.55}$$

and, by (2.44),

$$\gamma_{\|\mathbf{u}\oplus_{\mathrm{M}}\|\mathbf{v}\|} = \gamma_{\mathbf{u}}\gamma_{\mathbf{v}}\left(1 + \frac{\|\mathbf{u}\|\|\mathbf{v}\|}{s^2}\right) \qquad (2.56)$$

for all $\mathbf{u}, \mathbf{v} \in \mathbb{V}_s$. The restricted Möbius addition in (2.54) and (2.55) is both commutative and associative. The one in (2.55) gives rise to the Möbius *gyrotriangle inequality* in the following theorem.

Theorem 2.17.   (The Möbius Gyrotriangle Inequality).

$$\|\mathbf{u}\oplus\mathbf{v}\| \leq \|\mathbf{u}\|\oplus\|\mathbf{v}\| \qquad (2.57)$$

for all $\mathbf{u}, \mathbf{v}$ *in a Möbius gyrogroup* $(\mathbb{V}_s, \oplus)$.

Proof. By (2.56) and (2.44), with $\oplus_{\mathrm{M}} = \oplus$, and by the Cauchy-Schwarz inequality [28] we have

$$
\begin{aligned}
\gamma_{\|\mathbf{u}\|\oplus\|\mathbf{v}\|} &= \gamma_{\mathbf{u}}\gamma_{\mathbf{v}}\left(1 + \frac{\|\mathbf{u}\|\|\mathbf{v}\|}{s^2}\right)\\
&= \gamma_{\mathbf{u}}\gamma_{\mathbf{v}}\sqrt{1 + \frac{2}{s^2}\|\mathbf{u}\|\|\mathbf{v}\| + \frac{1}{s^4}\|\mathbf{u}\|^2\|\mathbf{v}\|^2}\\
&\geq \gamma_{\mathbf{u}}\gamma_{\mathbf{v}}\sqrt{1 + \frac{2}{s^2}\mathbf{u}\cdot\mathbf{v} + \frac{1}{s^4}\|\mathbf{u}\|^2\|\mathbf{v}\|^2}\\
&= \gamma_{\mathbf{u}\oplus\mathbf{v}}\\
&= \gamma_{\|\mathbf{u}\oplus\mathbf{v}\|}
\end{aligned}
\qquad (2.58)
$$

for all $\mathbf{u}, \mathbf{v}$ in the Möbius gyrogroup $(\mathbb{V}_s, \oplus)$. But $\gamma_{\mathbf{x}} = \gamma_{\|\mathbf{x}\|}, \mathbf{x} \in \mathbb{V}_s$, is a monotonically increasing function of $\|\mathbf{x}\|, 0 \leq \|\mathbf{x}\| < s$. Hence, (2.58) implies

$$\|\mathbf{u}\oplus\mathbf{v}\| \leq \|\mathbf{u}\|\oplus\|\mathbf{v}\| \qquad (2.59)$$

for all $\mathbf{u}, \mathbf{v}$ in any Möbius gyrogroup $(\mathbb{V}_s, \oplus)$. $\square$

2.3   EINSTEIN GYROGROUPS

Attempts to measure the absolute velocity of the earth through the hypothetical ether had failed. The most famous of these experiments is one performed by Michelson and Morley in 1887 [13]. It was 18 years later before the null results of these experiments were finally explained by Einstein in terms of a new velocity addition law that bears his name, which he introduced in his 1905 paper that founded the special theory of relativity [9, 10].

Contrasting Newtonian velocities, which are vectors in the Euclidean three-space $\mathbb{R}^3$, Einsteinian velocities must be relativistically admissible, that is, their magnitude must not exceed the vacuum speed of light, which is about $3 \times 10^5\ km\cdot sec^{-1}$.

Let c be the vacuum speed of light, and let

$$\mathbb{R}_c^3 = \{\mathbf{v} \in \mathbb{R}^3 : \|\mathbf{v}\| < c\} \tag{2.60}$$

be the c-ball of all relativistically admissible velocities of material particles. It is the open ball of radius c, centered at the origin of the Euclidean three-space $\mathbb{R}^3$, consisting of all vectors $\mathbf{v}$ in $\mathbb{R}^3$ with magnitude $\|\mathbf{v}\|$ smaller than c. Einstein addition $\oplus_E$ in the c-ball is given by the equation

$$\mathbf{u} \oplus_E \mathbf{v} = \frac{1}{1 + \frac{\mathbf{u} \cdot \mathbf{v}}{c^2}} \left\{ \mathbf{u} + \mathbf{v} + \frac{1}{c^2} \frac{\gamma_\mathbf{u}}{1 + \gamma_\mathbf{u}} (\mathbf{u} \times (\mathbf{u} \times \mathbf{v})) \right\} \tag{2.61}$$

for all $\mathbf{u}, \mathbf{v} \in \mathbb{R}_c^3$, where $\mathbf{u} \cdot \mathbf{v}$ is the inner product that the ball $\mathbb{R}_c^3$ inherits from its space $\mathbb{R}^3$, $\mathbf{u} \times \mathbf{v}$ is the vector product in $\mathbb{R}_c^3 \subset \mathbb{R}^3$, and where $\gamma_\mathbf{u}$ is the gamma factor

$$\gamma_\mathbf{u} = \frac{1}{\sqrt{1 - \frac{\|\mathbf{u}\|^2}{c^2}}} \tag{2.62}$$

in the c-ball.

Owing to the vector identity,

$$(\mathbf{x} \times \mathbf{y}) \times \mathbf{z} = -(\mathbf{y} \cdot \mathbf{z})\mathbf{x} + (\mathbf{x} \cdot \mathbf{z})\mathbf{y} \tag{2.63}$$

$\mathbf{x}, \mathbf{y}, \mathbf{z} \in \mathbb{R}^3$, that holds in $\mathbb{R}^3$, Einstein addition (2.61) can also be written in the form

$$\mathbf{u} \oplus_E \mathbf{v} = \frac{1}{1 + \frac{\mathbf{u} \cdot \mathbf{v}}{c^2}} \left\{ \mathbf{u} + \frac{1}{\gamma_\mathbf{u}} \mathbf{v} + \frac{1}{c^2} \frac{\gamma_\mathbf{u}}{1 + \gamma_\mathbf{u}} (\mathbf{u} \cdot \mathbf{v})\mathbf{u} \right\} \tag{2.64}$$

[54], which remains valid in higher dimensions.

Einstein addition (2.64) of relativistically admissible velocities was introduced by Einstein in his 1905 paper [10, p. 141] where the magnitudes of the two sides of Einstein addition (2.64) are presented. One has to remember here that the Euclidean 3-vector algebra was not so widely known in 1905 and, consequently, was not used by Einstein. Einstein calculated in [9] the behavior of the velocity components parallel and orthogonal to the relative velocity between inertial systems, which is as close as one can get without vectors to the vectorial version (2.64).

Example 2.18.   (The Components of Einstein Addition in $\mathbb{R}_c^3$). Let us introduce a Cartesian coordinate system Σ with coordinates labeled x, y, and z into $\mathbb{R}^3$ so that each point of the ball $\mathbb{R}_c^3$ is represented by its coordinates $(x, y, z)^t$ (exponent t denotes transposition) relative to Σ, satisfying the condition $x^2 + y^2 + z^2 < c^2$. Each point $(x, y, z)^t$ of the ball is identified with the vector from the origin $(0, 0, 0)^t$ of Σ, which is the center of the ball, to the point $(x, y, z)^t$. Furthermore, let $\mathbf{u}, \mathbf{v}, \mathbf{w} \in \mathbb{R}_c^3$ be three vectors in $\mathbb{R}_c^3 \subset \mathbb{R}^3$ given by their components relative to Σ,

$$\mathbf{u} = \begin{pmatrix} u_1 \\ u_2 \\ u_3 \end{pmatrix}, \qquad \mathbf{v} = \begin{pmatrix} v_1 \\ v_2 \\ v_3 \end{pmatrix}, \qquad \mathbf{w} = \begin{pmatrix} w_1 \\ w_2 \\ w_3 \end{pmatrix}, \tag{2.65}$$

where

$$\mathbf{w} = \mathbf{u} \oplus_{\mathrm{E}} \mathbf{v}. \tag{2.66}$$

The dot product of $\mathbf{u}$ and $\mathbf{v}$ is given by the equation

$$\mathbf{u} \cdot \mathbf{v} = u_1 v_1 + u_2 v_2 + u_3 v_3 \tag{2.67}$$

and the squared norm $\|\mathbf{u}\|^2 = \mathbf{u} \cdot \mathbf{u} =: \mathbf{u}^2$ of $\mathbf{u}$ is given by the equation

$$\|\mathbf{u}\|^2 = u_1^2 + u_2^2 + u_3^2. \tag{2.68}$$

Hence, it follows from the coordinate independent vector representation (2.64) of Einstein addition that the coordinate dependent Einstein addition (2.66) relative to the Cartesian coordinate system Σ takes the form

$$\begin{pmatrix} w_1 \\ w_2 \\ w_3 \end{pmatrix} = \frac{1}{1 + \dfrac{u_1 v_1 + u_2 v_2 + u_3 v_3}{c^2}}$$

$$\times \left\{ \left[1 + \frac{1}{c^2} \frac{\gamma_{\mathbf{u}}}{1 + \gamma_{\mathbf{u}}} (u_1 v_1 + u_2 v_2 + u_3 v_3) \right] \begin{pmatrix} u_1 \\ u_2 \\ u_3 \end{pmatrix} + \frac{1}{\gamma_{\mathbf{u}}} \begin{pmatrix} v_1 \\ v_2 \\ v_3 \end{pmatrix} \right\}, \tag{2.69}$$

where

$$\gamma_{\mathbf{u}} = \frac{1}{\sqrt{1 - \dfrac{u_1^2 + u_2^2 + u_3^2}{c^2}}}. \tag{2.70}$$

The x, y, and z components of Einstein addition (2.66) are w_1, w_2, and w_3, respectively, in (2.69). For two-dimensional illustrations of Einstein addition (2.69) one may impose the condition $u_3 = v_3 = 0$, implying $w_3 = 0$.

In this book vector equations and identities are represented in a coordinate free form, like Einstein addition in (2.64). These, however, must be converted into a coordinate dependent form, like the components of Einstein addition in (2.69), in numerical and graphical illustrations, as illustrated in Figs. 3.1–3.2, p. 80.

In the Newtonian limit, $c \to \infty$, the ball $\mathbb{R}_c^3$ of all relativistically admissible velocities expands to the whole of its space $\mathbb{R}^3$, as we see from (2.60), and Einstein addition $\oplus_{\mathrm{E}}$ in $\mathbb{R}_c^3$ reduces to the ordinary vector addition $+$ in $\mathbb{R}^3$, as we see from (2.64) and (2.62).

Suggestively, we extend Einstein addition of relativistically admissible velocities by abstraction in the following definition of Einstein addition in the ball.

Definition 2.19.   (Einstein Addition in the Ball). Let $\mathbb{V}$ be a real inner product space and let $\mathbb{V}_s$ be the s-ball of $\mathbb{V}$,

$$\mathbb{V}_s = \{ \mathbf{v} \in \mathbb{V} : \|\mathbf{v}\| < s \}, \tag{2.71}$$

where $s > 0$ is an arbitrarily fixed constant. Einstein addition $\oplus_E$ is a binary operation in $\mathbb{V}_s$ given by the equation

$$\mathbf{u} \oplus_E \mathbf{v} = \frac{1}{1 + \frac{\mathbf{u} \cdot \mathbf{v}}{s^2}} \left\{ \mathbf{u} + \frac{1}{\gamma_{\mathbf{u}}} \mathbf{v} + \frac{1}{s^2} \frac{\gamma_{\mathbf{u}}}{1 + \gamma_{\mathbf{u}}} (\mathbf{u} \cdot \mathbf{v}) \mathbf{u} \right\}, \tag{2.72}$$

where $\gamma_{\mathbf{u}}$ is the gamma factor

$$\gamma_{\mathbf{u}} = \frac{1}{\sqrt{1 - \frac{\|\mathbf{u}\|^2}{s^2}}} \tag{2.73}$$

in the s-ball $\mathbb{V}_s$, and where $\cdot$ and $\|\cdot\|$ are the inner product and norm that the ball $\mathbb{V}_s$ inherits from its space $\mathbb{V}$.

Like Möbius addition in the ball, one can show by computer algebra that Einstein addition in the ball is a gyrocommutative gyrogroup operation, giving rise to the Einstein ball gyrogroup $(\mathbb{V}_s, \oplus_E)$.

Einstein addition satisfies the mutually equivalent gamma identities

$$\gamma_{\mathbf{u} \oplus_E \mathbf{v}} = \gamma_{\mathbf{u}} \gamma_{\mathbf{v}} \left(1 + \frac{\mathbf{u} \cdot \mathbf{v}}{s^2} \right) \tag{2.74}$$

and

$$\gamma_{\ominus_E \mathbf{u} \oplus_E \mathbf{v}} = \gamma_{\mathbf{u} \ominus_E \mathbf{v}} = \gamma_{\mathbf{u}} \gamma_{\mathbf{v}} \left(1 - \frac{\mathbf{u} \cdot \mathbf{v}}{s^2} \right) \tag{2.75}$$

for all $\mathbf{u}, \mathbf{v} \in \mathbb{V}_s$.

When the nonzero vectors $\mathbf{u}$ and $\mathbf{v}$ in the ball $\mathbb{V}_s$ of $\mathbb{V}$ are parallel in $\mathbb{V}$, $\mathbf{u} \| \mathbf{v}$, that is, $\mathbf{u} = \lambda \mathbf{v}$ for some $0 \neq \lambda \in \mathbb{R}$, Einstein addition reduces to the Einstein addition of parallel velocities,

$$\mathbf{u} \boxplus_E \mathbf{v} = \mathbf{u} \oplus_E \mathbf{v} = \frac{\mathbf{u} + \mathbf{v}}{1 + \frac{1}{s^2} \|\mathbf{u}\| \|\mathbf{v}\|}, \qquad \mathbf{u} \| \mathbf{v}. \tag{2.76}$$

Hence, accordingly,

$$\|\mathbf{u}\| \boxplus_E \|\mathbf{v}\| = \|\mathbf{u}\| \oplus_E \|\mathbf{v}\| = \frac{\|\mathbf{u}\| + \|\mathbf{v}\|}{1 + \frac{1}{s^2} \|\mathbf{u}\| \|\mathbf{v}\|} \tag{2.77}$$

for all $\mathbf{u}, \mathbf{v} \in \mathbb{V}_s$. Remarkably, Einstein addition and Möbius addition of parallel vectors coincide, as we see from (2.76) and (2.54).

The restricted Einstein addition in (2.76) and (2.77) is both commutative and associative. Accordingly, the restricted Einstein addition is a group operation, as Einstein noted in [9]; see [10, p. 142]. In contrast, Einstein made no remark about group properties of his addition of velocities that need not be parallel. Indeed, the general Einstein addition is not a group operation but, rather, a gyrocommutative gyrogroup operation, a structure that was discovered more than 80 years later, in 1988 [52].

Einstein addition obeys the *gyrotriangle inequality*, presented in the following theorem.

Theorem 2.20.   (The Einstein Gyrotriangle Inequality).

$$\|\mathbf{u}\oplus_{E}\mathbf{v}\| \leq \|\mathbf{u}\|\oplus_{E}\|\mathbf{v}\| \tag{2.78}$$

for all $\mathbf{u}, \mathbf{v}$ *in an Einstein gyrogroup* $(\mathbb{V}_s, \oplus_E)$.

Proof. By (2.74), with $\oplus_E = \oplus$, and by the Cauchy-Schwarz inequality [28], we have

$$\begin{aligned}
\gamma_{\|\mathbf{u}\|\oplus\|\mathbf{v}\|} &= \gamma_{\mathbf{u}}\gamma_{\mathbf{v}}\left(1 + \frac{\|\mathbf{u}\|\,\|\mathbf{v}\|}{s^2}\right) \\
&\geq \gamma_{\mathbf{u}}\gamma_{\mathbf{v}}\left(1 + \frac{\mathbf{u}\cdot\mathbf{v}}{s^2}\right) \\
&= \gamma_{\mathbf{u}\oplus\mathbf{v}} \\
&= \gamma_{\|\mathbf{u}\oplus\mathbf{v}\|}
\end{aligned} \tag{2.79}$$

for all $\mathbf{u}, \mathbf{v}$ in the Einstein gyrogroup $(\mathbb{V}_s, \oplus_E)$. But $\gamma_{\mathbf{x}} = \gamma_{\|\mathbf{x}\|}, \mathbf{x} \in \mathbb{V}_s$, is a monotonically increasing function of $\|\mathbf{x}\|, 0 \leq \|\mathbf{x}\| < s$. Hence, by (2.79),

$$\|\mathbf{u}\oplus_{E}\mathbf{v}\| \leq \|\mathbf{u}\|\oplus_{E}\|\mathbf{v}\| \tag{2.80}$$

for all $\mathbf{u}, \mathbf{v}$ in any Einstein gyrogroup $(\mathbb{V}_s, \oplus_E)$. $\square$

Einstein gyrations

$$\text{gyr}[\mathbf{u}, \mathbf{v}] : \mathbb{V}_s \to \mathbb{V}_s \tag{2.81}$$

are automorphisms of the Einstein gyrogroup $(\mathbb{V}_s, \oplus_E)$,

$$\text{gyr}[\mathbf{u}, \mathbf{v}] \in Aut(\mathbb{V}_s, \oplus_E), \tag{2.82}$$

given by the gyrator identity, Theorem 1.13(10), p. 11,

$$\text{gyr}[\mathbf{u}, \mathbf{v}]\mathbf{w} = \ominus_{E}(\mathbf{u}\oplus_{E}\mathbf{v})\oplus_{E}\{\mathbf{u}\oplus_{E}(\mathbf{v}\oplus_{E}\mathbf{w})\}, \tag{2.83}$$

and are further specified by the equation

$$\text{gyr}[\mathbf{u}, \mathbf{v}]\mathbf{w} = \mathbf{w} + \frac{A\mathbf{u} + B\mathbf{v}}{D} \tag{2.84}$$

where

$$A = -\frac{1}{s^2}\frac{\gamma_{\mathbf{u}}^2}{(\gamma_{\mathbf{u}}+1)}(\gamma_{\mathbf{v}}-1)(\mathbf{u}\cdot\mathbf{w}) + \frac{1}{s^2}\gamma_{\mathbf{u}}\gamma_{\mathbf{v}}(\mathbf{v}\cdot\mathbf{w})$$

$$+ \frac{2}{s^4}\frac{\gamma_{\mathbf{u}}^2\gamma_{\mathbf{v}}^2}{(\gamma_{\mathbf{u}}+1)(\gamma_{\mathbf{v}}+1)}(\mathbf{u}\cdot\mathbf{v})(\mathbf{v}\cdot\mathbf{w})$$

$$(2.85)$$

$$B = -\frac{1}{s^2}\frac{\gamma_{\mathbf{v}}}{\gamma_{\mathbf{v}}+1}\{\gamma_{\mathbf{u}}(\gamma_{\mathbf{v}}+1)(\mathbf{u}\cdot\mathbf{w}) + (\gamma_{\mathbf{u}}-1)\gamma_{\mathbf{v}}(\mathbf{v}\cdot\mathbf{w})\}$$

$$D = \gamma_{\mathbf{u}}\gamma_{\mathbf{v}}(1 + \frac{\mathbf{u}\cdot\mathbf{v}}{s^2}) + 1 = \gamma_{\mathbf{u}\oplus\mathbf{v}} + 1 > 0$$

for all $\mathbf{u}, \mathbf{v}, \mathbf{w} \in \mathbb{V}_s$. Allowing $\mathbf{w} \in \mathbb{V} \supset \mathbb{V}_s$ in (2.84)–(2.85), gyrations gyr$[\mathbf{u}, \mathbf{v}]$ are expendable to linear maps of $\mathbb{V}$ for all $\mathbf{u}, \mathbf{v} \in \mathbb{V}_s$.

In each of the three special cases when (i) $\mathbf{u} = \mathbf{0}$, or (ii) $\mathbf{v} = \mathbf{0}$, or (iii) $\mathbf{u}$ and $\mathbf{v}$ are parallel in $\mathbb{V}$, $\mathbf{u}\|\mathbf{v}$, we have $A\mathbf{u} + B\mathbf{v} = \mathbf{0}$ so that gyr$[\mathbf{u}, \mathbf{v}]$ is trivial,

$$\begin{aligned} \text{gyr}[\mathbf{0}, \mathbf{v}]\mathbf{w} &= \mathbf{w} \\ \text{gyr}[\mathbf{u}, \mathbf{0}]\mathbf{w} &= \mathbf{w} \\ \text{gyr}[\mathbf{u}, \mathbf{v}]\mathbf{w} &= \mathbf{w}, \qquad \mathbf{u}\|\mathbf{v} \end{aligned}$$

$$(2.86)$$

for all $\mathbf{u}, \mathbf{v} \in \mathbb{V}_s$ and all $\mathbf{w} \in \mathbb{V}$.

In a 3-dimensional Einstein gyrovector space, gyrations turn out to coincide with the relativistic effect known as Thomas precession, [61, Sec. 10.3]. Being a concrete example of the abstract gyration, Thomas precession emerges clearly and naturally in the study of Einstein's special theory of relativity in terms of its underlying hyperbolic geometry. In contrast, the study of Thomas precession in the literature involves some confusion, as pointed out in [27].

Gyrations preserve the inner product that the ball $\mathbb{V}_s$ inherits from its real inner product space $\mathbb{V}$,

$$\text{gyr}[\mathbf{u}, \mathbf{v}]\mathbf{a}\cdot\text{gyr}[\mathbf{u}, \mathbf{v}]\mathbf{b} = \mathbf{a}\cdot\mathbf{b}$$

$$(2.87)$$

for all $\mathbf{a}, \mathbf{b}, \mathbf{u}, \mathbf{v}, \mathbf{w} \in \mathbb{V}_s$. Hence, in particular, Einstein gyrations preserve the norm that the ball $\mathbb{V}_s$ inherits from its real inner product space $\mathbb{V}$ so that, by the gyrocommutative law,

$$\|\mathbf{u}\oplus_{\scriptscriptstyle E}\mathbf{v}\| = \|\text{gyr}[\mathbf{u}, \mathbf{v}](\mathbf{v}\oplus_{\scriptscriptstyle E}\mathbf{u})\| = \|\mathbf{v}\oplus_{\scriptscriptstyle E}\mathbf{u}\| \,.$$

$$(2.88)$$

Einstein gyrogroup cooperation (1.20), p. 7, in an Einstein gyrogroup $(\mathbb{V}_s, \oplus_{\text{E}})$ is given by Einstein coaddition, $\boxplus_{\text{E}}$,

$$
\begin{aligned}
\mathbf{u} \boxplus_{\text{E}} \mathbf{v} &= \frac{\gamma_{\mathbf{u}} + \gamma_{\mathbf{v}}}{\gamma_{\mathbf{u}}^2 + \gamma_{\mathbf{v}}^2 + \gamma_{\mathbf{u}} \gamma_{\mathbf{v}} \left(1 + \frac{\mathbf{u} \cdot \mathbf{v}}{s^2}\right) - 1} (\gamma_{\mathbf{u}} \mathbf{u} + \gamma_{\mathbf{v}} \mathbf{v}) \\
&= \frac{\gamma_{\mathbf{u}} + \gamma_{\mathbf{v}}}{(\gamma_{\mathbf{u}} + \gamma_{\mathbf{v}})^2 - (\gamma_{\mathbf{u} \ominus_{\text{E}} \mathbf{v}} + 1)} (\gamma_{\mathbf{u}} \mathbf{u} + \gamma_{\mathbf{v}} \mathbf{v}) \\
&= 2 \otimes_{\text{E}} \frac{\gamma_{\mathbf{u}} \mathbf{u} + \gamma_{\mathbf{v}} \mathbf{v}}{\gamma_{\mathbf{u}} + \gamma_{\mathbf{v}}},
\end{aligned}
\tag{2.89}
$$

where the scalar multiplication by the factor 2 is defined by the equation $2 \otimes_{\text{E}} \mathbf{v} = \mathbf{v} \oplus_{\text{E}} \mathbf{v}$. A more general definition of the scalar multiplication by any real number will be presented and studied in Chap. 3.

The following equivalent form of (2.89),

$$
\mathbf{u} \boxplus_{\text{E}} \mathbf{v} = \frac{\gamma_{\mathbf{u}} \mathbf{u} + \gamma_{\mathbf{v}} \mathbf{v}}{\gamma_{\mathbf{u}} + \gamma_{\mathbf{v}} - \dfrac{\gamma_{\ominus_{\text{E}} \mathbf{u} \oplus_{\text{E}} \mathbf{v}} + 1}{\gamma_{\mathbf{u}} + \gamma_{\mathbf{v}}}},
\tag{2.90}
$$

proves useful in its geometrically guided extension to more than two summands in [61, Eqs. (10.66)–(10.67), p. 425].

Einstein coaddition is commutative, as we see from (2.89), and as expected from Theorem 2.3, p. 34, and it satisfies the gamma identity, [61, Eq. (3.197), p. 93],

$$
\begin{aligned}
\gamma_{\mathbf{u} \boxplus_{\text{E}} \mathbf{v}} &= \frac{\gamma_{\mathbf{u}}^2 + \gamma_{\mathbf{v}}^2 + \gamma_{\mathbf{u}} \gamma_{\mathbf{v}} \left(1 + \frac{\mathbf{u} \cdot \mathbf{v}}{s^2}\right) - 1}{\gamma_{\mathbf{u}} \gamma_{\mathbf{v}} \left(1 - \frac{\mathbf{u} \cdot \mathbf{v}}{s^2}\right) + 1} \\
&= \frac{(\gamma_{\mathbf{u}} + \gamma_{\mathbf{v}})^2}{\gamma_{\mathbf{u} \ominus_{\text{E}} \mathbf{v}} + 1} - 1.
\end{aligned}
\tag{2.91}
$$

Other interesting identities that Einstein addition possesses are, [61, Eq. (3.198)],

$$
\gamma_{\mathbf{u} \oplus_{\text{E}} \mathbf{v}} (\mathbf{u} \oplus_{\text{E}} \mathbf{v}) = \frac{\gamma_{\mathbf{v}} + \gamma_{\mathbf{u} \oplus_{\text{E}} \mathbf{v}}}{1 + \gamma_{\mathbf{u}}} \gamma_{\mathbf{u}} \mathbf{u} + \gamma_{\mathbf{v}} \mathbf{v}
\tag{2.92}
$$

and, [61, Eq. (3.199)],

$$
\gamma_{\mathbf{u} \boxplus_{\text{E}} \mathbf{v}} (\mathbf{u} \boxplus_{\text{E}} \mathbf{v}) = \frac{\gamma_{\mathbf{u}} + \gamma_{\mathbf{v}}}{1 + \gamma_{\mathbf{u} \ominus_{\text{E}} \mathbf{v}}} (\gamma_{\mathbf{u}} \mathbf{u} + \gamma_{\mathbf{v}} \mathbf{v}).
\tag{2.93}
$$

2.4   GYROGROUP ISOMORPHISM

In full analogy with group isomorphism, a gyrogroup isomorphism is a bijective map (that is, one-to-one, onto) of a gyrogroup onto another gyrogroup that preserves gyrogroup operations. Two gyrogroups that are related by a gyrogroup isomorphism are said to be isomorphic. We will find in this section that Einstein and Möbius gyrogroups are isomorphic.

For any element $\mathbf{v}$ of a gyrogroup $(G, \oplus)$ we use the notation

$$2\otimes\mathbf{v} = \mathbf{v}\oplus\mathbf{v}\,, \tag{2.94}$$

so that if $\mathbf{v}$ is an element of an Einstein gyrogroup $(\mathbb{V}_s, \oplus_{\mathrm{E}})$ we have $2\otimes\mathbf{v} = \mathbf{v}\oplus_{\mathrm{E}}\mathbf{v}$, and if $\mathbf{v}$ is an element of a Möbius gyrogroup $(\mathbb{V}_s, \oplus_{\mathrm{M}})$ we have $2\otimes\mathbf{v} = \mathbf{v}\oplus_{\mathrm{M}}\mathbf{v}$.

The unique solution to the equation $2\otimes\mathbf{v} = \mathbf{u}$ for the unknown $\mathbf{v}$ in an Einstein gyrogroup $(\mathbb{V}_s, \oplus_{\mathrm{E}})$ is the *Einstein half*, $\mathbf{u} = (1/2)\otimes\mathbf{v}$, given by the equation

$$\tfrac{1}{2}\otimes\mathbf{v} = \frac{\gamma_{\mathbf{v}}}{1 + \gamma_{\mathbf{v}}}\mathbf{v}\,. \tag{2.95}$$

As expected, Einstein half satisfies the identity

$$
\begin{aligned}
2\otimes(\tfrac{1}{2}\otimes\mathbf{v}) &= 2\otimes\frac{\gamma_{\mathbf{v}}}{1 + \gamma_{\mathbf{v}}}\mathbf{v} \\
&= \frac{\gamma_{\mathbf{v}}}{1 + \gamma_{\mathbf{v}}}\mathbf{v} \oplus_{\mathrm{E}} \frac{\gamma_{\mathbf{v}}}{1 + \gamma_{\mathbf{v}}}\mathbf{v} \\
&= \mathbf{v}\,.
\end{aligned}
\tag{2.96}
$$

Möbius addition, $\oplus_{\mathrm{M}}$, and Einstein addition, $\oplus_{\mathrm{E}}$, are related by the two mutually equivalent identities

$$
\begin{aligned}
\mathbf{u}_e\oplus_{\mathrm{E}}\mathbf{v}_e &= 2\otimes(\tfrac{1}{2}\otimes\mathbf{u}_e\oplus_{\mathrm{M}}\tfrac{1}{2}\otimes\mathbf{v}_e) \\
\mathbf{u}_m\oplus_{\mathrm{M}}\mathbf{v}_m &= \tfrac{1}{2}\otimes(2\otimes\mathbf{u}_m\oplus_{\mathrm{E}}2\otimes\mathbf{v}_m)
\end{aligned}
\tag{2.97}
$$

for all $\mathbf{u}_e, \mathbf{v}_e \in (\mathbb{V}_s, \oplus_{\mathrm{E}})$, and all $\mathbf{u}_m, \mathbf{v}_m \in (\mathbb{V}_s, \oplus_{\mathrm{M}})$.

Identities (2.97) suggest the map

$$\phi_{\mathrm{EM}} : (\mathbb{V}_s, \oplus_{\mathrm{M}}) \;\rightarrow\; (\mathbb{V}_s, \oplus_{\mathrm{E}}), \qquad \mathbf{v}_m \mapsto \mathbf{v}_e = 2\otimes\mathbf{v}_m\,, \tag{2.98}$$

or simply,

$$\phi_{\mathrm{EM}}\mathbf{v}_m = 2\otimes\mathbf{v}_m\,, \tag{2.99}$$

and its inverse map

$$\phi_{\mathrm{ME}} : (\mathbb{V}_s, \oplus_{\mathrm{E}}) \;\rightarrow\; (\mathbb{V}_s, \oplus_{\mathrm{M}}), \qquad \mathbf{v}_e \mapsto \mathbf{v}_m = \tfrac{1}{2}\otimes\mathbf{v}_e\,, \tag{2.100}$$

or simply,

$$\phi_{ME}\mathbf{v}_e = \tfrac{1}{2}\otimes\mathbf{v}_e . \tag{2.101}$$

Being inverse to each other, the gyrogroup maps ϕ_{EM} and ϕ_{ME} are bijective (that is, maps which are one-to-one and onto). Moreover, they preserve their respective gyrogroup operations. Indeed, by (2.99) and the second identity in (2.97) we have

$$\begin{aligned}
\phi_{EM}(\mathbf{u}_m\oplus_{M}\mathbf{v}_m) &= 2\otimes(\mathbf{u}_m\oplus_{M}\mathbf{v}_m)\\
&= 2\otimes\{\tfrac{1}{2}\otimes(2\otimes\mathbf{u}_m\oplus_{E}2\otimes\mathbf{v}_m)\}\\
&= 2\otimes\mathbf{u}_m\oplus_{E}2\otimes\mathbf{v}_m\\
&= \phi_{EM}\mathbf{u}_m\oplus_{E}\phi_{EM}\mathbf{v}_m ,
\end{aligned} \tag{2.102}$$

noting that $2\otimes$ and $\tfrac{1}{2}\otimes$ play against each other to their mutual annihilation, as in (2.96). Similarly, by (2.101) and the first identity in (2.97) we have

$$\begin{aligned}
\phi_{ME}(\mathbf{u}_e\oplus_{E}\mathbf{v}_e) &= \tfrac{1}{2}\otimes(\mathbf{u}_e\oplus_{E}\mathbf{v}_e)\\
&= \tfrac{1}{2}\otimes\{2\otimes(\tfrac{1}{2}\otimes\mathbf{u}_e\oplus_{M}\tfrac{1}{2}\mathbf{v}_e)\}\\
&= \tfrac{1}{2}\otimes\mathbf{u}_e\oplus_{M}\tfrac{1}{2}\otimes\mathbf{v}_e\\
&= \phi_{ME}\mathbf{u}_e\oplus_{M}\phi_{ME}\mathbf{v}_e .
\end{aligned} \tag{2.103}$$

As such, the maps ϕ_{EM} and ϕ_{ME} are isomorphisms that establish gyrogroup isomorphism between any Möbius gyrogroup $(\mathbb{V}_s, \oplus_{M})$ and its corresponding Einstein gyrogroup $(\mathbb{V}_s, \oplus_{E})$.

By definition, gyrogroup isomorphisms preserve gyrogroup operations. Additionally, they preserve gyrations and gyrogroup cooperations as well, as indicated in the following theorem, for Einstein and Möbius gyrogroups.

Theorem 2.21. *Let $(\mathbb{V}_s, \oplus_{M})$ and $(\mathbb{V}_s, \oplus_{E})$ be Möbius and Einstein gyrogroups in the s-ball $\mathbb{V}_s$ of a real inner product space, $\mathbb{V}$, and let gyr_{M}, $\boxplus_{M}$, and gyr_{E}, $\boxplus_{E}$, be their gyrators and cooperations, respectively. Then,*

$$\begin{aligned}
\phi_{EM}\mathrm{gyr}_{M}[\mathbf{u}_m, \mathbf{v}_m]\mathbf{w}_m &= \mathrm{gyr}_{E}[\phi_{EM}\mathbf{u}_m, \phi_{EM}\mathbf{v}_m]\phi_{EM}\mathbf{w}_m\\
\phi_{EM}(\mathbf{u}_m \boxplus_{M} \mathbf{v}_m) &= \phi_{EM}\mathbf{u}_m \boxplus_{E} \phi_{EM}\mathbf{v}_m ,
\end{aligned} \tag{2.104}$$

and, similarly,

$$\begin{aligned}
\phi_{ME}\mathrm{gyr}_{E}[\mathbf{u}_e, \mathbf{v}_e]\mathbf{w}_e &= \mathrm{gyr}_{M}[\phi_{ME}\mathbf{u}_e, \phi_{ME}\mathbf{v}_e]\phi_{ME}\mathbf{w}_e\\
\phi_{ME}(\mathbf{u}_e \boxplus_{E} \mathbf{v}_e) &= \phi_{ME}\mathbf{u}_e \boxplus_{M} \phi_{ME}\mathbf{v}_e .
\end{aligned} \tag{2.105}$$

Proof. The first identity in (2.104) results from the following chain of equations:

$$
\phi_{EM}\mathrm{gyr}_{M}[\mathbf{u}_m,\mathbf{v}_m]\mathbf{w}_m \overset{(1)}{=\!=} \phi_{EM}[\ominus_{M}(\mathbf{u}_m\oplus_{M}\mathbf{v}_m)\oplus_{M}\{\mathbf{u}_m\oplus_{M}(\mathbf{v}_m\oplus_{M}\mathbf{w}_m)\}]
$$
$$
\overset{(2)}{=\!=} \ominus_{E}(\phi_{EM}\mathbf{u}_m\oplus_{E}\phi_{EM}\mathbf{v}_m)\oplus_{E}\{\phi_{EM}\mathbf{u}_m\oplus_{E}(\phi_{EM}\mathbf{v}_m\oplus_{E}\phi_{EM}\mathbf{w}_m)\}
$$
$$
\overset{(3)}{=\!=} \mathrm{gyr}_{E}[\phi_{EM}\mathbf{u}_m,\phi_{EM}\mathbf{v}_m]\phi_{EM}\mathbf{w}_m , \tag{2.106}
$$

for all $\mathbf{u}_m,\mathbf{v}_m,\mathbf{w}_m \in (\mathbb{V}_s,\oplus_{M})$. Derivation of the numbered equalities in (2.106) follows.

(1) Follows from the gyrator identity in Theorem 1.13(10), p. 11.

(2) Follows from (1) since the isomorphism ϕ_{EM} takes Möbius addition into Einstein addition according to (2.102).

(3) Follows from (2) by the gyrator identity in Theorem 1.13(10).

The second identity in (2.104) results from the following chain of equations:

$$
\phi_{EM}(\mathbf{u}_m \boxplus_{M} \mathbf{v}_m) \overset{(1)}{=\!=} \phi_{EM}(\mathbf{u}_m\oplus_{M}\mathrm{gyr}_{M}[\mathbf{u}_m.\ominus_{M}\mathbf{v}_m]\mathbf{v}_m)
$$
$$
\overset{(2)}{=\!=} \phi_{EM}\mathbf{u}_m\oplus_{E}\phi_{EM}\mathrm{gyr}_{M}[\mathbf{u}_m,\ominus_{M}\mathbf{v}_m]\mathbf{v}_m
$$
$$
\overset{(3)}{=\!=} \phi_{EM}\mathbf{u}_m\oplus_{E}\mathrm{gyr}_{E}[\phi_{EM}\mathbf{u}_m,\ominus_{E}\phi_{EM}\mathbf{v}_m]\phi_{EM}\mathbf{v}_m
$$
$$
\overset{(4)}{=\!=} \phi_{EM}\mathbf{u}_m \boxplus_{E} \phi_{EM}\mathbf{v}_m , \tag{2.107}
$$

for all $\mathbf{u}_m,\mathbf{v}_m \in (\mathbb{V}_s,\oplus_{M})$. Derivation of the numbered equalities in (2.107) follows.

(1) This equation results from Def. 1.9, Chap. 1, p. 7, of the gyrogroup cooperation $\boxplus$.

(2) Follows from (1) since the isomorphism ϕ_{EM} takes Möbius addition into Einstein addition according to (2.102).

(3) Follows from (2) since the isomorphism ϕ_{EM} preserves gyrations according to the first identity in (2.104).

(4) Follows from (3) by Def. 1.9, Chap. 1, of the gyrogroup cooperation $\boxplus$.

Finally, the proof of (2.105) is similar to that of (2.104).   □

2.5   EXERCISES

(1) Verify the Möbius gamma identity (2.44).

(2) Verify the Möbius coaddition formula (2.52).

(3) Verify Identity (2.53).

(4) Verify Identity (2.74).

(5) Verify Identity (2.76).

(6) Verify Identity (2.87).

(7) Verify Identities $(2.89) - (2.93)$.

(8) Verify Identity (2.96).

(9) Verify Identities (2.97).

(10) Employ (2.98) to prove the identities

$$\gamma_{\mathbf{v}_e} = \gamma_{2\otimes_M \mathbf{v}_m} = 2\gamma^2_{\mathbf{v}_m} - 1 \tag{2.108}$$

and

$$\gamma_{\mathbf{v}_e} \mathbf{v}_e = \gamma_{2\otimes_M \mathbf{v}_m} 2\otimes_M \mathbf{v}_m = 2\gamma^2_{\mathbf{v}_m} \mathbf{v}_m . \tag{2.109}$$

Hint: See $(3.127) - (3.128)$, p. 89. See also [61, p. 208]. See also (3.111), p. 82.

CHAPTER 3

Gyrovector Spaces

Some gyrocommutative gyrogroups admit scalar multiplication, giving rise to gyrovector spaces. The latter, in turn, are analogous to vector spaces just as gyrogroups are analogous to groups and gyrocommutative gyrogroups are analogous to commutative groups. Indeed, we will find in this book that gyrovector spaces provide the algebraic setting for hyperbolic geometry just as vector spaces provide the algebraic setting for Euclidean geometry.

3.1 DEFINITION AND FIRST GYROVECTOR SPACE THEOREMS

We have already encountered real inner product spaces in the definition of Möbius gyrogroups, Def. 2.16, p. 41, and in the definition of Einstein gyrogroups, Def. 2.19, p. 46. We have also encountered the Cauchy-Schwarz inequality in the proof of Theorem 2.17, p. 43, and Theorem 2.20, p. 47. It is now the time to present these formally, in order to pave the way for the definition of gyrovector spaces.

Definition 3.1. (Real Inner Product Vector Spaces). *A real inner product vector space* $(\mathbb{V}, +, \cdot)$ *(vector space, in short) is a real vector space together with a map*

$$\mathbb{V} \times \mathbb{V} \to \mathbb{R}, \qquad (\mathbf{u}, \mathbf{v}) \mapsto \mathbf{u}\cdot\mathbf{v} \tag{3.1}$$

called a real inner product, satisfying the following properties for all $\mathbf{u}, \mathbf{v}, \mathbf{w} \in \mathbb{V}$ *and* $r \in \mathbb{R}$:

(1) $\mathbf{v}\cdot\mathbf{v} \geq 0$, *with equality if, and only if,* $\mathbf{v} = 0$
(2) $\mathbf{u}\cdot\mathbf{v} = \mathbf{v}\cdot\mathbf{u}$
(3) $(\mathbf{u} + \mathbf{v})\cdot\mathbf{w} = \mathbf{u}\cdot\mathbf{w} + \mathbf{v}\cdot\mathbf{w}$
(4) $(r\mathbf{u})\cdot\mathbf{v} = r(\mathbf{u}\cdot\mathbf{v})$

The norm $\|\mathbf{v}\|$ *of* $\mathbf{v} \in \mathbb{V}$ *is given by the equation* $\|\mathbf{v}\|^2 = \mathbf{v}\cdot\mathbf{v}$.

Note that the properties of vector spaces imply (i) the Cauchy-Schwarz inequality

$$|\mathbf{u}\cdot\mathbf{v}| \leq \|\mathbf{u}\|\|\mathbf{v}\| \tag{3.2}$$

for all $\mathbf{u}, \mathbf{v} \in \mathbb{V}$; and (ii) the *positive definiteness* of the inner product, according to which $\mathbf{u}\cdot\mathbf{v} = 0$ for all $\mathbf{u} \in \mathbb{V}$ implies $\mathbf{v} = 0$ [28].

Definition 3.2. (Real Inner Product Gyrovector Spaces). *A real inner product gyrovector space* $(G, \oplus, \otimes)$ *(gyrovector space, in short) is a gyrocommutative gyrogroup* $(G, \oplus)$ *that obeys the following axioms:*

(1) G is a subset of a real inner product vector space $\mathbb{V}$ called the carrier of G, $G \subset \mathbb{V}$, from which it inherits its inner product, $\cdot$, and norm, $\|\cdot\|$, which are invariant under gyroautomorphisms, that is:

(V1) $\text{gyr}[\mathbf{u}, \mathbf{v}]\mathbf{a} \cdot \text{gyr}[\mathbf{u}, \mathbf{v}]\mathbf{b} = \mathbf{a} \cdot \mathbf{b}$ *Inner Product Gyroinvariance*

for all points $\mathbf{a}, \mathbf{b}, \mathbf{u}, \mathbf{v} \in G$.

(2) G admits a scalar multiplication, $\otimes$, possessing the following properties. For all real numbers $r, r_1, r_2 \in \mathbb{R}$ and all points $\mathbf{a} \in G$:

(V2) $1 \otimes \mathbf{a} = \mathbf{a}$ *Identity Scalar Multiplication*

(V3) $(r_1 + r_2) \otimes \mathbf{a} = r_1 \otimes \mathbf{a} \oplus r_2 \otimes \mathbf{a}$ *Scalar Distributive Law*

(V4) $(r_1 r_2) \otimes \mathbf{a} = r_1 \otimes (r_2 \otimes \mathbf{a})$ *Scalar Associative Law*

(V5) $\dfrac{|r| \otimes \mathbf{a}}{\|r \otimes \mathbf{a}\|} = \dfrac{\mathbf{a}}{\|\mathbf{a}\|}, \quad \mathbf{a} \neq \mathbf{0},\ r \neq 0$ *Scaling Property*

(V6) $\text{gyr}[\mathbf{u}, \mathbf{v}](r \otimes \mathbf{a}) = r \otimes \text{gyr}[\mathbf{u}, \mathbf{v}]\mathbf{a}$ *Gyroautomorphism Property*

(V7) $\text{gyr}[r_1 \otimes \mathbf{v}, r_2 \otimes \mathbf{v}] = I$ *Identity (Trivial) Gyroautomorphism.*

(3) Real, one-dimensional vector space structure $(\|G\|, \oplus, \otimes)$ for the set $\|G\|$ of one-dimensional "vectors" (see, for instance, [ ])

(V8) $\|G\| = \{\pm\|\mathbf{a}\| : \mathbf{a} \in G\} \subset \mathbb{R}$ *Vector Space*

with vector addition $\oplus$ and scalar multiplication $\otimes$, such that for all $r \in \mathbb{R}$ and $\mathbf{a}, \mathbf{b} \in G$,

(V9) $\|r \otimes \mathbf{a}\| = |r| \otimes \|\mathbf{a}\|$ *Homogeneity Property*

(V10) $\|\mathbf{a} \oplus \mathbf{b}\| \leq \|\mathbf{a}\| \oplus \|\mathbf{b}\|$ *Gyrotriangle Inequality.*

Remark 3.3.   *We use the notation $(r_1 \otimes \mathbf{a}) \oplus (r_2 \otimes \mathbf{b}) = r_1 \otimes \mathbf{a} \oplus r_2 \otimes \mathbf{b}$, and $\mathbf{a} \otimes r = r \otimes \mathbf{a}$. Our ambiguous use of $\oplus$ and $\otimes$ in Def. 3.2 as interrelated operations in the gyrovector space $(G, \oplus, \otimes)$ and in its associated vector space $(\|G\|, \oplus, \otimes)$ should raise no confusion since the sets in which these operations operate are always clear from the context. These operations in the former (gyrovector space $(G, \oplus, \otimes)$) are nonassociative-nondistributive gyrovector space operations, and in the latter (vector space $(\|G\|, \oplus, \otimes)$) are associative-distributive vector space operations. Additionally, the gyro-addition $\oplus$ is gyrocommutative in the former and commutative in the latter. Note that in the vector space $(\|G\|, \oplus, \otimes)$ gyrations are trivial so that $\boxplus = \oplus$ in $\|G\|$.*

While the operations $\oplus$ and $\otimes$ have distinct interpretations in the gyrovector space G and in the vector space $\|G\|$, they are related to one another by the gyrovector space axioms $(V9)$ and $(V10)$. The analogies that conventions about the ambiguous use of $\oplus$ and $\otimes$ in G and $\|G\|$ share with similar vector space conventions are obvious. Indeed, in vector spaces we use the same notation, $+$, for the addition operation between vectors and between their magnitudes, and same notation for the scalar multiplication *between two scalars and between a scalar and a vector. In full analogy, in gyrovector spaces we use the same notation, $\oplus$, for the gyroaddition operation between gyrovectors and between their magnitudes, and the same notation, $\otimes$, for the scalar gyromultiplication between two scalars and between a scalar and a gyrovector.*

Immediate consequences of the gyrovector space axioms are presented in the following theorem.

Theorem 3.4. *Let $(G, \oplus, \otimes)$ be a gyrovector space whose carrier vector space is $\mathbb{V}$, and let 0, $\mathbf{0}$ and $\mathbf{0}_{\mathbb{V}}$ be the neutral elements of the real line $(\mathbb{R}, +)$, the gyrocommutative gyrogroup $(G, \oplus)$, and the vector space $(\mathbb{V}, +)$, respectively. Then, for all $n \in \mathbb{N}, r \in \mathbb{R}$, and $\mathbf{a} \in G$:*

(1) $0 \otimes \mathbf{a} = \mathbf{0}$.

(2) $n \otimes \mathbf{a} = \mathbf{a} \oplus \ \ldots \ \oplus \mathbf{a}$ \qquad *(n terms).*

(3) $(-r) \otimes \mathbf{a} = \ominus(r \otimes \mathbf{a}) =: \ominus r \otimes \mathbf{a}$.

(4) $r \otimes \mathbf{0} = \mathbf{0}$.

(5) $r \otimes (\ominus \mathbf{a}) = \ominus(r \otimes \mathbf{a}) =: \ominus r \otimes \mathbf{a}$.

(6) $\|\ominus \mathbf{a}\| = \|\mathbf{a}\|$.

(7) $\mathbf{0} = \mathbf{0}_{\mathbb{V}}$ *(The neutral elements of $G \subset \mathbb{V}$ and $\mathbb{V}$ are equal).*

(8) $r \otimes \mathbf{a} = \mathbf{0} \iff (r = 0 \text{ or } \mathbf{a} = \mathbf{0})$.

Proof. (1) Follows from the scalar distributive law $(V3)$,

$$r \otimes \mathbf{a} = (r + 0) \otimes \mathbf{a} = r \otimes \mathbf{a} \oplus 0 \otimes \mathbf{a}, \tag{3.3}$$

so that, by a left cancellation, $0 \otimes \mathbf{a} = \ominus(r \otimes \mathbf{a}) \oplus (r \otimes \mathbf{a}) = \mathbf{0}$.

(2) Follows from $(V2)$, and the scalar distributive law $(V3)$. Indeed, with "..." signifying "n terms", we have

$$\mathbf{a} \oplus \ldots \oplus \mathbf{a} = 1 \otimes \mathbf{a} \oplus \ldots \oplus 1 \otimes \mathbf{a} = (1 + \ldots + 1) \otimes \mathbf{a} = n \otimes \mathbf{a}. \tag{3.4}$$

(3) Results from (1) and the scalar distributive law $(V3)$,

$$\mathbf{0} = 0 \otimes \mathbf{a} = (r - r) \otimes \mathbf{a} = r \otimes \mathbf{a} \oplus (-r) \otimes \mathbf{a}, \tag{3.5}$$

implying $\ominus(r \otimes \mathbf{a}) = (-r) \otimes \mathbf{a}$.

(4) Follows from (1), $(V4)$, $(V3)$, (3),

$$
\begin{aligned}
r\otimes\mathbf{0} &= r\otimes(0\otimes\mathbf{a}) \\
&= r\otimes((1-1)\otimes\mathbf{a}) \\
&= (r(1-1))\otimes\mathbf{a} \\
&= (r-r)\otimes\mathbf{a} \\
&= r\otimes\mathbf{a}\oplus(-r)\otimes\mathbf{a} \\
&= r\otimes\mathbf{a}\oplus(\ominus(r\otimes\mathbf{a})) \\
&= r\otimes\mathbf{a}\ominus r\otimes\mathbf{a} \\
&= \mathbf{0}\,.
\end{aligned} \tag{3.6}
$$

(5) Follows from the scalar distributive law and (4) above,

$$
r\otimes\mathbf{a}\oplus r\otimes(\ominus\mathbf{a}) = r\otimes(\mathbf{a}\ominus\mathbf{a}) = r\otimes\mathbf{0} = \mathbf{0}\,. \tag{3.7}
$$

(6) Follows from (3), the homogeneity property $(V9)$, and $(V2)$,

$$
\|\ominus\mathbf{a}\| = \|(-1)\otimes\mathbf{a}\| = |-1|\otimes\|\mathbf{a}\| = 1\otimes\|\mathbf{a}\| = \|1\otimes\mathbf{a}\| = \|\mathbf{a}\|\,. \tag{3.8}
$$

(7) Results from (4), $(V9)$, and $(V8)$ as follows.

$$
\|\mathbf{0}\| = \|2\otimes\mathbf{0}\| = 2\otimes\|\mathbf{0}\| = \|\mathbf{0}\|\oplus\|\mathbf{0}\|\,, \tag{3.9}
$$

implying $\|\mathbf{0}\| = \|\mathbf{0}\|\ominus\|\mathbf{0}\| = 0$ in the vector space $(\|G\|, \oplus, \otimes)$. This equation, $\|\mathbf{0}\| = 0$, is valid in the vector space $\mathbb{V}$ as well, where it implies $\mathbf{0} = \mathbf{0}_{\mathbb{V}}$.

(8) Results from the following considerations. Suppose $r\otimes\mathbf{a} = \mathbf{0}$, but $r \neq 0$. Then, by $(V1)$, $(V4)$, and (4) we have

$$
\mathbf{a} = 1\otimes\mathbf{a} = (1/r)\otimes(r\otimes\mathbf{a}) = (1/r)\otimes\mathbf{0} = \mathbf{0}\,. \tag{3.10}
$$
□

In the special case when all the gyrations of a gyrovector space are trivial, the gyrovector space reduces to a vector space.

In general, gyroaddition does not distribute with scalar multiplication,

$$
r\otimes(\mathbf{a}\oplus\mathbf{b}) \neq r\otimes\mathbf{a}\oplus r\otimes\mathbf{b}\,. \tag{3.11}
$$

However, gyrovector spaces possess a weak distributive law, called the monodistributive law, presented in the following theorem.

Theorem 3.5.   (The Monodistributive Law). *A gyrovector space* $(G, \oplus, \otimes)$ *possesses the monodistributive law*

$$
r\otimes(r_1\otimes\mathbf{a}\oplus r_2\otimes\mathbf{a}) = r\otimes(r_1\otimes\mathbf{a})\oplus r\otimes(r_2\otimes\mathbf{a}) \tag{3.12}
$$

for all $r, r_1, r_2 \in \mathbb{R}$, *and* $\mathbf{a}\in G$.

Proof. The proof follows from $(V3)$ and $(V4)$,

$$
\begin{aligned}
r\otimes(r_1\otimes\mathbf{a}\oplus r_2\otimes\mathbf{a}) &= r\otimes\{(r_1+r_2)\otimes\mathbf{a}\} \\
&= (r(r_1+r_2))\otimes\mathbf{a} \\
&= (rr_1+rr_2)\otimes\mathbf{a} \\
&= (rr_1)\otimes\mathbf{a}\oplus(rr_2)\otimes\mathbf{a} \\
&= r\otimes(r_1\otimes\mathbf{a})\oplus r\otimes(r_2\otimes\mathbf{a})\,.
\end{aligned}
\tag{3.13}
$$

□

Along with the gyrotriangle equality that gyrovector spaces obey by Axiom (V10), they obey the *reverse gyrotriangle inequality* as well. The reverse gyrotriangle inequality in gyrovector spaces follows directly from the gyrotriangle inequality itself, as we see in the following theorem and its proof.

Theorem 3.6.   (The Reverse Gyrotriangle Inequality). *Any two elements, $\mathbf{a}$, $\mathbf{b}$, of a gyrovector space* $(G, \oplus, \otimes)$ *obey the reverse gyrotriangle law*

$$
|\,\|\mathbf{a}\|\ominus\|\mathbf{b}\|\,| \le \|\mathbf{a}\ominus\mathbf{b}\|\,.
\tag{3.14}
$$

Proof. Let us suppose without loss of generality that $\|\mathbf{a}\|$ is no smaller than $\|\mathbf{b}\|$ (otherwise, we interchange the roles of $\mathbf{a}$ and $\mathbf{b}$). By the left gyroassociative law we have

$$
\begin{aligned}
\mathbf{a} &= \mathbf{a}\oplus\mathbf{0} \\
&= \mathbf{a}\oplus(\ominus\mathbf{b}\oplus\mathbf{b}) \\
&= (\mathbf{a}\ominus\mathbf{b})\oplus\mathrm{gyr}[\mathbf{a},\ominus\mathbf{b}]\mathbf{b}\,,
\end{aligned}
\tag{3.15}
$$

so that by the gyrotriangle inequality $(V10)$ in Def. 3.2 of gyrovector spaces we have

$$
\begin{aligned}
\|\mathbf{a}\| &= \|(\mathbf{a}\ominus\mathbf{b})\oplus\mathrm{gyr}[\mathbf{a},\ominus\mathbf{b}]\mathbf{b}\| \\
&\le \|\mathbf{a}\ominus\mathbf{b}\|\oplus\|\mathrm{gyr}[\mathbf{a},\ominus\mathbf{b}]\mathbf{b}\| \\
&= \|\mathbf{a}\ominus\mathbf{b}\|\oplus\|\mathbf{b}\|\,,
\end{aligned}
\tag{3.16}
$$

noting that gyrations preserve the norm according to Axiom $(V1)$ of gyrovector spaces in Def. 3.2.

The binary operation $\oplus$ in the extreme right-hand side of (3.16) is a commutative and associative binary operation in the one-dimensional vector space $(\|G\|, \oplus, \otimes)$, according to Axiom $(V8)$ of gyrovector spaces in Def. 3.2, as noted in Remark 3.3. Hence, (3.16) implies

$$
\|\mathbf{a}\|\ominus\|\mathbf{b}\| \le \|\mathbf{a}\ominus\mathbf{b}\|\,,
\tag{3.17}
$$

thus verifying (3.14).

□

Definition 3.7.   (Gyrovector Space Automorphisms). *An automorphism τ of a gyrovector space $(G, \oplus, \otimes)$, $\tau \in Aut(G, \oplus, \otimes)$, is a bijective self-map of G, $\tau : \ G \to G$, which preserves its structure, that is, (i) binary operation, (ii) scalar multiplication, and (iii) inner product,*

$$\tau(\mathbf{a} \oplus \mathbf{b}) = \tau \mathbf{a} \oplus \tau \mathbf{b}$$
$$\tau(r \otimes \mathbf{a}) = r \otimes \tau \mathbf{a} \tag{3.18}$$
$$\tau \mathbf{a} \cdot \tau \mathbf{b} = \mathbf{a} \cdot \mathbf{b} \, .$$

The automorphisms of the gyrovector space $(G, \oplus, \otimes)$ form a group denoted $Aut(G, \oplus, \otimes)$, with group operation given by automorphism composition.

Clearly, it follows from Def. 3.7 that gyrations are special automorphisms. Therefore, gyrations are also called gyroautomorphisms when one wishes to emphasize this result. While the set of all automorphisms of a gyrovector space forms a group under automorphism composition, in general, the set of all gyroautomorphisms of a gyrovector space does not form a group under gyroautomorphism composition.

Definition 3.8.   (Motions of Gyrovector Spaces). *The motions of a gyrovector space $(G, \oplus, \otimes)$ are all its left gyrotranslations λ_x, $x \in G$, (1.69), p. 19, and its automorphisms $\tau \in Aut(G, \oplus, \otimes)$.*

Since scalar multiplication in a gyrovector space does not distribute with the gyrovector space gyroaddition, the *Two-Sum Identity* in the following theorem proves useful.

Theorem 3.9.   (The Two-Sum Identity). *Let $(G, \oplus, \otimes)$ be a gyrovector space. Then*

$$2 \otimes (\mathbf{a} \oplus \mathbf{b}) = \mathbf{a} \oplus (2 \otimes \mathbf{b} \oplus \mathbf{a})$$
$$= \mathbf{a} \boxplus (\mathbf{a} \oplus 2 \otimes \mathbf{b}) \tag{3.19}$$

for any $\mathbf{a}, \mathbf{b} \in G$.

Proof. Employing the right gyroassociative law, the identity gyr$[\mathbf{b}, \mathbf{b}] = I$, the left gyroassociative law, and the gyrocommutative law we have the following chain of equations that gives the first identity in (3.19):

$$\begin{aligned}
\mathbf{a} \oplus (2 \otimes \mathbf{b} \oplus \mathbf{a}) &= \mathbf{a} \oplus ((\mathbf{b} \oplus \mathbf{b}) \oplus \mathbf{a}) \\
&= \mathbf{a} \oplus (\mathbf{b} \oplus (\mathbf{b} \oplus \text{gyr}[\mathbf{b}, \mathbf{b}]\mathbf{a})) \\
&= \mathbf{a} \oplus (\mathbf{b} \oplus (\mathbf{b} \oplus \mathbf{a})) \\
&= (\mathbf{a} \oplus \mathbf{b}) \oplus \text{gyr}[\mathbf{a}, \mathbf{b}](\mathbf{b} \oplus \mathbf{a}) \\
&= (\mathbf{a} \oplus \mathbf{b}) \oplus (\mathbf{a} \oplus \mathbf{b}) \\
&= 2 \otimes (\mathbf{a} \oplus \mathbf{b}) \, .
\end{aligned} \tag{3.20}$$

The second equality in (3.19) follows from the first one and Theorem 2.10, p. 37.   □

A gyrovector space is a gyrometric space with a gyrodistance function that obeys the gyrotriangle inequality. Remarkably, the gyrodistance function of gyrovector spaces and its associated gyrotriangle inequality that gyrovector spaces obey are fully analogous to their classical counterparts, that is, to the distance function of vector spaces and its associated triangle inequality that vector spaces obey.

Definition 3.10. (The Gyrodistance Function, Gyrometric). *Let $G = (G, \oplus, \otimes)$ be a gyrovector space. Its gyrometric is given by the gyrodistance function $d_\oplus(\mathbf{a}, \mathbf{b}) : G \times G \to \mathbb{R}^{\geq 0} := \{r \in \mathbb{R} : r \geq 0\}$,*

$$d_\oplus(\mathbf{a}, \mathbf{b}) = \|\ominus\mathbf{a}\oplus\mathbf{b}\| = \|\mathbf{b}\ominus\mathbf{a}\|, \tag{3.21}$$

where $d_\oplus(\mathbf{a}, \mathbf{b})$ is the gyrodistance of $\mathbf{a}$ to $\mathbf{b}$.

By Def. 3.2 of gyrovector spaces, gyroautomorphisms preserve the inner product. Hence, they are isometries, that is, they preserve the norm as well. Accordingly, the identity $\|\ominus\mathbf{a}\oplus\mathbf{b}\| = \|\mathbf{b}\ominus\mathbf{a}\|$ in Def. 3.10 follows from the gyrocommutative law,

$$\|\ominus\mathbf{a}\oplus\mathbf{b}\| = \|\text{gyr}[\ominus\mathbf{a}, \mathbf{b}](\mathbf{b}\ominus\mathbf{a})\| = \|\mathbf{b}\ominus\mathbf{a}\|. \tag{3.22}$$

Theorem 3.11. (The Gyrotriangle Inequality). *The gyrometric of a gyrovector space $(G, \oplus, \otimes)$ satisfies the gyrotriangle inequality*

$$\|\ominus\mathbf{a}\oplus\mathbf{c}\| \leq \|\ominus\mathbf{a}\oplus\mathbf{b}\|\oplus\|\ominus\mathbf{b}\oplus\mathbf{c}\| \tag{3.23}$$

for all $\mathbf{a}, \mathbf{b}, \mathbf{c} \in G$.

Proof. By Theorem 1.18, p. 16, we have,

$$\ominus\mathbf{a}\oplus\mathbf{c} = (\ominus\mathbf{a}\oplus\mathbf{b})\oplus\text{gyr}[\ominus\mathbf{a}, \mathbf{b}](\ominus\mathbf{b}\oplus\mathbf{c}). \tag{3.24}$$

Hence, by the gyrotriangle inequality $(V10)$ in Def. 3.2 we have

$$\begin{aligned}
\|\ominus\mathbf{a}\oplus\mathbf{c}\| &= \|(\ominus\mathbf{a}\oplus\mathbf{b})\oplus\text{gyr}[\ominus\mathbf{a}, \mathbf{b}](\ominus\mathbf{b}\oplus\mathbf{c})\| \\
&\leq \|\ominus\mathbf{a}\oplus\mathbf{b}\|\oplus\|\text{gyr}[\ominus\mathbf{a}, \mathbf{b}](\ominus\mathbf{b}\oplus\mathbf{c})\| \\
&= \|\ominus\mathbf{a}\oplus\mathbf{b}\|\oplus\|\ominus\mathbf{b}\oplus\mathbf{c}\|.
\end{aligned} \tag{3.25}$$

□

The basic properties of the gyrodistance function $d_\oplus$ are:

(1) $d_\oplus(\mathbf{a}, \mathbf{b}) \geq 0$

(2) $d_\oplus(\mathbf{a}, \mathbf{b}) = 0$ if and only if $\mathbf{a} = \mathbf{b}$.

(3) $d_\oplus(\mathbf{a}, \mathbf{b}) = d_\oplus(\mathbf{b}, \mathbf{a})$

(4) $d_\oplus(\mathbf{a}, \mathbf{c}) \leq d_\oplus(\mathbf{a}, \mathbf{b}) \oplus d_\oplus(\mathbf{b}, \mathbf{c})$ (gyrotriangle inequality),

$\mathbf{a}, \mathbf{b}, \mathbf{c} \in G$.

Theorem 3.12.   *The gyrodistance in a gyrovector space is invariant under automorphisms and left gyro-translations.*

Proof. By Def. 3.7, automorphisms $\tau \in Aut(G, \oplus, \otimes)$ preserve the inner product. As such, they preserve the norm and, hence, the gyrodistance,

$$\|\tau\mathbf{b} \ominus \tau\mathbf{a}\| = \|\tau(\mathbf{b} \ominus \mathbf{a})\| = \|\mathbf{b} \ominus \mathbf{a}\|, \tag{3.26}$$

for all $\mathbf{a}, \mathbf{b}$ in the gyrovector space $(G, \oplus, \otimes)$. Hence, the gyrodistance is invariant under automorphisms.

Let $\mathbf{a}, \mathbf{b}, \mathbf{x} \in G$ be any three points in a gyrovector space $(G, \oplus, \otimes)$, and let the points $\mathbf{a}$ and $\mathbf{b}$ be left gyrotranslated by $\mathbf{x}$ into $\mathbf{a}'$ and $\mathbf{b}'$, respectively,

$$\begin{aligned} \mathbf{a}' &= \mathbf{x} \oplus \mathbf{a} \\ \mathbf{b}' &= \mathbf{x} \oplus \mathbf{b} . \end{aligned} \tag{3.27}$$

Then, by the Gyrotranslation Theorem 2.11, p. 37, we have

$$\mathbf{b}' \ominus \mathbf{a}' = (\mathbf{x} \oplus \mathbf{b}) \ominus (\mathbf{x} \oplus \mathbf{a}) = \mathrm{gyr}[\mathbf{x}, \mathbf{b}](\mathbf{b} \ominus \mathbf{a}), \tag{3.28}$$

so that

$$\|\mathbf{b}' \ominus \mathbf{a}'\| = \|\mathrm{gyr}[\mathbf{x}, \mathbf{b}](\mathbf{b} \ominus \mathbf{a})\| = \|\mathbf{b} \ominus \mathbf{a}\|. \tag{3.29}$$

Hence, the gyrodistance is invariant under left gyrotranslations.   □

3.2   GYROLINES

Definition 3.13.   (Gyrolines, Gyrosegments). *Let* $\mathbf{a}, \mathbf{b}$ *be any two distinct points in a gyrovector space* $(G, \oplus, \otimes)$. *The gyroline in G that passes through the points* $\mathbf{a}$ *and* $\mathbf{b}$ *is the set of all points*

$$L = \mathbf{a} \oplus (\ominus \mathbf{a} \oplus \mathbf{b}) \otimes t \tag{3.30}$$

in G with $t \in \mathbb{R}$. *The gyrovector space expression in* (3.30) *is called the representation of the gyroline L in terms of the two points* $\mathbf{a}$ *and* $\mathbf{b}$ *that it contains.*

A gyroline segment (or, a gyrosegment) $\mathbf{ab}$ *with endpoints* $\mathbf{a}$ *and* $\mathbf{b}$ *is the set of all points in* (3.30) *with* $0 \leq t \leq 1$. *The gyrolength* $|\mathbf{ab}|$ *of the gyrosegment* $\mathbf{ab}$ *is the gyrodistance between* $\mathbf{a}$ *and* $\mathbf{b}$,

$$|\mathbf{ab}| = d_\oplus(\mathbf{a}, \mathbf{b}) = \|\ominus \mathbf{a} \oplus \mathbf{b}\|. \tag{3.31}$$

Two gyrosegments are congruent if they have the same gyrolength.

Considering the real parameter t as "time", the gyroline (3.30) passes through the point **a** at time $t = 0$ and, owing to the left cancellation law, it passes through the point **b** at time $t = 1$.

It is anticipated in Def. 3.13 that the gyroline is uniquely represented by any two given points that it contains. The following theorem shows that this is indeed the case.

Theorem 3.14.   *Two gyrolines that share two distinct points are coincident.*

Proof. Let

$$\mathbf{a}\oplus(\ominus\mathbf{a}\oplus\mathbf{b})\otimes t \tag{3.32}$$

be a gyroline that contains two given distinct points, $\mathbf{p}_1$ and $\mathbf{p}_2$, in a gyrovector space $(G, \oplus, \otimes)$. Then, there exist real numbers $t_1, t_2 \in \mathbb{R}, t_1 \neq t_2$, such that

$$\begin{aligned}\mathbf{p}_1 &= \mathbf{a}\oplus(\ominus\mathbf{a}\oplus\mathbf{b})\otimes t_1 \\ \mathbf{p}_2 &= \mathbf{a}\oplus(\ominus\mathbf{a}\oplus\mathbf{b})\otimes t_2 \,.\end{aligned} \tag{3.33}$$

A gyroline containing the points $\mathbf{p}_1$ and $\mathbf{p}_2$ has the form

$$\mathbf{p}_1\oplus(\ominus\mathbf{p}_1\oplus\mathbf{p}_2)\otimes t \,, \tag{3.34}$$

which, by means of (3.33) is reducible to (3.32) with a reparametrization.

Indeed, we manipulate the following chain of equations

$$\begin{aligned}
&\mathbf{p}_1\oplus(\ominus\mathbf{p}_1\oplus\mathbf{p}_2)\otimes t \\
&\overset{(1)}{=\!=\!=} [\mathbf{a}\oplus(\ominus\mathbf{a}\oplus\mathbf{b})\otimes t_1]\oplus\{\ominus[\mathbf{a}\oplus(\ominus\mathbf{a}\oplus\mathbf{b})\otimes t_1]\oplus[\mathbf{a}\oplus(\ominus\mathbf{a}\oplus\mathbf{b})\otimes t_2]\}\otimes t \\
&\overset{(2)}{=\!=\!=} [\mathbf{a}\oplus(\ominus\mathbf{a}\oplus\mathbf{b})\otimes t_1]\oplus\mathrm{gyr}[\mathbf{a}, (\ominus\mathbf{a}\oplus\mathbf{b})\otimes t_1]\{\ominus(\ominus\mathbf{a}\oplus\mathbf{b})\otimes t_1\oplus(\ominus\mathbf{a}\oplus\mathbf{b})\otimes t_2\}\otimes t \\
&\overset{(3)}{=\!=\!=} [\mathbf{a}\oplus(\ominus\mathbf{a}\oplus\mathbf{b})\otimes t_1]\oplus\mathrm{gyr}[\mathbf{a}, (\ominus\mathbf{a}\oplus\mathbf{b})\otimes t_1]\{(\ominus\mathbf{a}\oplus\mathbf{b})\otimes(-t_1 + t_2)\}\otimes t \\
&\overset{(4)}{=\!=\!=} [\mathbf{a}\oplus(\ominus\mathbf{a}\oplus\mathbf{b})\otimes t_1]\oplus\mathrm{gyr}[\mathbf{a}, (\ominus\mathbf{a}\oplus\mathbf{b})\otimes t_1](\ominus\mathbf{a}\oplus\mathbf{b})\otimes((-t_1 + t_2)t) \\
&\overset{(5)}{=\!=\!=} \mathbf{a}\oplus\{(\ominus\mathbf{a}\oplus\mathbf{b})\otimes t_1\oplus(\ominus\mathbf{a}\oplus\mathbf{b})\otimes((-t_1 + t_2)t)\} \\
&\overset{(6)}{=\!=\!=} \mathbf{a}\oplus(\ominus\mathbf{a}\oplus\mathbf{b})\otimes(t_1 + (-t_1 + t_2)t) \,,
\end{aligned} \tag{3.35}$$

obtaining the gyroline (3.32) with a reparametrization. It is a reparametrization in which the original gyroline parameter t in (3.32) is replaced by the new gyroline parameter $t_1 + (-t_1 + t_2)t$, with $t_2 - t_1 \neq 0$. Derivation of the numbered equalities in (3.35) follows.

(1) This equation is obtained by a substitution of $\mathbf{p}_1$ and $\mathbf{p}_2$ from (3.33).

(2) Follows from (1) by the Gyrotranslation Theorem 2.11, p. 37.

(3) Follows from (2) by the scalar distributive law.

(4) Follows from (3) by the scalar associative law.

(5) Follows from (4) by the left gyroassociative law. Indeed, the application of the left gyroassociative law to (5) results in (4).

(6) Follows from (5) by the scalar distributive law.

Hence, by (3.35), any gyroline (3.32) that contains the two distinct points $\mathbf{p}_1$ and $\mathbf{p}_2$ coincides with the gyroline (3.34). □

Theorem 3.15. *A left gyrotranslation of a gyroline is, again, a gyroline.*

Proof. Let

$$L = \mathbf{a} \oplus (\ominus \mathbf{a} \oplus \mathbf{b}) \otimes t \tag{3.36}$$

be a gyroline L represented by its two points $\mathbf{a}$ and $\mathbf{b}$ in a gyrovector space $(G, \oplus, \otimes)$. The left gyrotranslation, $\mathbf{x} \oplus L$, of the gyroline L by any $\mathbf{x} \in G$ is given by the equation

$$\mathbf{x} \oplus L = \mathbf{x} \oplus \{\mathbf{a} \oplus (\ominus \mathbf{a} \oplus \mathbf{b}) \otimes t\}, \tag{3.37}$$

which can be recast in the form of a gyroline as shown in the following chain of equations:

$$
\begin{aligned}
\mathbf{x} \oplus L &\overset{(1)}{=\!=\!=} \mathbf{x} \oplus \{\mathbf{a} \oplus (\ominus \mathbf{a} \oplus \mathbf{b}) \otimes t\} \\
&\overset{(2)}{=\!=\!=} (\mathbf{x} \oplus \mathbf{a}) \oplus \mathrm{gyr}[\mathbf{x}, \mathbf{a}]\{(\ominus \mathbf{a} \oplus \mathbf{b}) \otimes t\} \\
&\overset{(3)}{=\!=\!=} (\mathbf{x} \oplus \mathbf{a}) \oplus \{\mathrm{gyr}[\mathbf{x}, \mathbf{a}](\ominus \mathbf{a} \oplus \mathbf{b})\} \otimes t \\
&\overset{(4)}{=\!=\!=} (\mathbf{x} \oplus \mathbf{a}) \oplus \{\ominus (\mathbf{x} \oplus \mathbf{a}) \oplus (\mathbf{x} \oplus \mathbf{b})\} \otimes t,
\end{aligned} \tag{3.38}
$$

thus obtaining a gyroline representation, (3.32), for the left gyrotranslated gyroline, $\mathbf{x} \oplus L$. Derivation of the numbered equalities in (3.38) follows.

(1) This equation gives the left gyrotranslated gyroline in (3.37).

(2) Follows from (1) by the left gyroassociative law.

(3) Follows from (2) by Axiom ($V6$) of gyrovector spaces.

(4) Follows from (3) by the Gyrotranslation Theorem 2.11, p. 37. □

Definition 3.16. (Gyrocollinearity). *Three points,* $\mathbf{a}_1, \mathbf{a}_2, \mathbf{a}_3,$ *in a gyrovector space* $(G, \oplus, \otimes)$ *are gyrocollinear if they lie on the same gyroline, that is, if there exist* $\mathbf{a}, \mathbf{b} \in G$ *such that*

$$\mathbf{a}_k = \mathbf{a} \oplus (\ominus \mathbf{a} \oplus \mathbf{b}) \otimes t_k \qquad (3.39)$$

for some $t_k \in \mathbb{R}$, $k = 1, 2, 3$. *Similarly, n points in G, n > 3, are gyrocollinear if any three of these points are gyrocollinear.*

The geometric concept of one point being between two others is an extremely important, yet at the same time, an extremely intuitive idea. As noted by Millman and Parker [32, p. 47], the concept of betweenness does not appear formally in Euclid, which leads to some logical flaws as, for instance, the one mentioned in [18, Fig. 3.2, p. 60].

Our definition of betweenness results in the Gyrotriangle Equality Theorem 3.28, p. 74, of the gyrotriangle equality. In contrast, some authors prefer to adopt the gyrotriangle equality of Theorem 3.28 as the definition of betweenness as, for instant, in [32, p. 47].

Definition 3.17. (Betweenness). *A point* $\mathbf{a}_2$ *lies between the points* $\mathbf{a}_1$ *and* $\mathbf{a}_3$ *in a gyrovector space* $(G, \oplus, \otimes)$

(*i*) *if the points* $\mathbf{a}_1, \mathbf{a}_2, \mathbf{a}_3$ *are gyrocollinear, that is, they are related by the equations*

$$\mathbf{a}_k = \mathbf{a} \oplus (\ominus \mathbf{a} \oplus \mathbf{b}) \otimes t_k \ , \qquad (3.40)$$

$k = 1, 2, 3,$ *for some* $\mathbf{a}, \mathbf{b} \in G, \mathbf{a} \neq \mathbf{b},$ *and some* $t_k \in \mathbb{R}$, *and*

(*ii*) *if either* $t_1 < t_2 < t_3$ *or* $t_3 < t_2 < t_1$.

The proof of the following Lemma 3.18 leads to Lemma 3.19 that gives a necessary and sufficient condition for one point being between two others.

Lemma 3.18. *Three distinct points,* $\mathbf{a}_1, \mathbf{a}_2,$ *and* $\mathbf{a}_3,$ *in a gyrovector space* $(G, \oplus, \otimes)$ *are gyrocollinear if and only if any one of these points, say* $\mathbf{a}_2,$ *can be expressed in terms of the two other points by the equation*

$$\mathbf{a}_2 = \mathbf{a}_1 \oplus (\ominus \mathbf{a}_1 \oplus \mathbf{a}_3) \otimes t_0 \qquad (3.41)$$

for some $t_0 \in \mathbb{R}$.

Proof. If the points $\mathbf{a}_1, \mathbf{a}_2, \mathbf{a}_3$ are gyrocollinear, then there exist distinct points $\mathbf{a}, \mathbf{b} \in G$ and distinct real number t_k such that

$$\mathbf{a}_k = \mathbf{a} \oplus (\ominus \mathbf{a} \oplus \mathbf{b}) \otimes t_k \ , \qquad (3.42)$$

$k = 1, 2, 3.$

Let

$$t_0 = \frac{t_2 - t_1}{t_3 - t_1}.$$ (3.43)

Then, the following chain of equations verifies (3.41):

$\mathbf{a}_1 \oplus (\ominus \mathbf{a}_1 \oplus \mathbf{a}_3) \otimes t_0$

$\overset{(1)}{=\!=\!=} [\mathbf{a} \oplus (\ominus \mathbf{a} \oplus \mathbf{b}) \otimes t_1] \oplus \{\ominus [\mathbf{a} \oplus (\ominus \mathbf{a} \oplus \mathbf{b}) \otimes t_1] \oplus [\mathbf{a} \oplus (\ominus \mathbf{a} \oplus \mathbf{b}) \otimes t_3]\} \otimes t_0$

$\overset{(2)}{=\!=\!=} [\mathbf{a} \oplus (\ominus \mathbf{a} \oplus \mathbf{b}) \otimes t_1] \oplus \mathrm{gyr}[\mathbf{a}, (\ominus \mathbf{a} \oplus \mathbf{b}) \otimes t_1] \{\ominus (\ominus \mathbf{a} \oplus \mathbf{b}) \otimes t_1 \oplus (\ominus \mathbf{a} \oplus \mathbf{b}) \otimes t_3\} \otimes t_0$

$\overset{(3)}{=\!=\!=} [\mathbf{a} \oplus (\ominus \mathbf{a} \oplus \mathbf{b}) \otimes t_1] \oplus \mathrm{gyr}[\mathbf{a}, (\ominus \mathbf{a} \oplus \mathbf{b}) \otimes t_1] \{(\ominus \mathbf{a} \oplus \mathbf{b}) \otimes (-t_1 + t_3)\} \otimes t_0$

$\overset{(4)}{=\!=\!=} [\mathbf{a} \oplus (\ominus \mathbf{a} \oplus \mathbf{b}) \otimes t_1] \oplus \mathrm{gyr}[\mathbf{a}, (\ominus \mathbf{a} \oplus \mathbf{b}) \otimes t_1] (\ominus \mathbf{a} \oplus \mathbf{b}) \otimes ((-t_1 + t_3) t_0)$ (3.44)

$\overset{(5)}{=\!=\!=} \mathbf{a} \oplus \{(\ominus \mathbf{a} \oplus \mathbf{b}) \otimes t_1 \oplus (\ominus \mathbf{a} \oplus \mathbf{b}) \otimes ((-t_1 + t_3) t_0)\}$

$\overset{(6)}{=\!=\!=} \mathbf{a} \oplus (\ominus \mathbf{a} \oplus \mathbf{b}) \otimes (t_1 + (-t_1 + t_3) t_0)$

$\overset{(7)}{=\!=\!=} \mathbf{a} \oplus (\ominus \mathbf{a} \oplus \mathbf{b}) \otimes t_2$

$\overset{(8)}{=\!=\!=} \mathbf{a}_2.$

Derivation of the numbered equalities in (3.38) follows.

(1) This equation is obtained by a substitution of $\mathbf{a}_1$ and $\mathbf{a}_3$ from (3.42).
(2) Follows from (1) by the Gyrotranslation Theorem 2.11, p. 37.
(3) Follows from (2) by Theorem 3.4(3), p. 57, and the scalar distributive law.
(4) Follows from (3) by the scalar associative law.
(5) Follows from (4) by the left gyroassociative law. Indeed, an application of the left gyroassociative law to (5) results in (4).
(6) Follows from (5) by the scalar distributive law.
(7) Follows from (6) by (3.43).
(8) Follows from (7) by (3.42).

Conversely, if (3.41) holds then the three points $\mathbf{a}_1$, $\mathbf{a}_2$, and $\mathbf{a}_3$ are gyrocollinear since, by (3.41), the point $\mathbf{a}_2$ lies on the gyroline that passes through the points $\mathbf{a}_1$ and $\mathbf{a}_3$. □

Lemma 3.19.   *A point $\mathbf{a}_2$ lies between the points $\mathbf{a}_1$ and $\mathbf{a}_3$ in a gyrovector space $(G, \oplus, \otimes)$ if and only if*

$$\mathbf{a}_2 = \mathbf{a}_1 \oplus (\ominus \mathbf{a}_1 \oplus \mathbf{a}_3) \otimes t_0$$ (3.45)

for some $0 < t_0 < 1$.

Proof. If $\mathbf{a}_2$ lies between $\mathbf{a}_1$ and $\mathbf{a}_3$, the points $\mathbf{a}_1, \mathbf{a}_2, \mathbf{a}_3$ are gyrocollinear by Def. 3.17, and there exist distinct points $\mathbf{a}, \mathbf{b} \in G$ and real numbers t_k such that

$$\mathbf{a}_k = \mathbf{a} \oplus (\ominus \mathbf{a} \oplus \mathbf{b}) \otimes t_k \,, \tag{3.46}$$

$k = 1, 2, 3$, and either $t_1 < t_2 < t_3$ or $t_3 < t_2 < t_1$.

Let

$$t_0 = \frac{t_2 - t_1}{t_3 - t_1} \,. \tag{3.47}$$

Then $0 < t_0 < 1$ and, following the chain of equations (3.44), we derive the desired identity

$$\mathbf{a}_1 \oplus (\ominus \mathbf{a}_1 \oplus \mathbf{a}_3) \otimes t_0 = \mathbf{a}_2 \,, \tag{3.48}$$

thus verifying (3.45) for $0 < t_0 < 1$.

Conversely, if (3.45) holds then, by Def. 3.17 with $t_1 = 0, t_2 = t_0$, and $t_3 = 1$, $\mathbf{a}_2$ lies between $\mathbf{a}_1$ and $\mathbf{a}_3$. $\qquad\square$

Lemma 3.20.   *The two equations*

$$\mathbf{b} = \mathbf{a} \oplus (\ominus \mathbf{a} \oplus \mathbf{c}) \otimes t \tag{3.49}$$

and

$$\mathbf{b} = \mathbf{c} \oplus (\ominus \mathbf{c} \oplus \mathbf{a}) \otimes (1 - t) \tag{3.50}$$

are equivalent for the parameter $t \in \mathbb{R}$ and all points $\mathbf{a}, \mathbf{b}, \mathbf{c}$ in a gyrovector space $(G, \oplus, \otimes)$.

Proof. The equivalence of (3.49) and (3.50) results from the extreme sides of the following chain of equation:

$$
\begin{aligned}
\mathbf{c} \oplus (\ominus \mathbf{c} \oplus \mathbf{a}) \otimes (1 - t) &\overset{(1)}{=\!=} \mathbf{c} \oplus \{(\ominus \mathbf{c} \oplus \mathbf{a}) \ominus (\ominus \mathbf{c} \oplus \mathbf{a}) \otimes t\} \\
&\overset{(2)}{=\!=} \{\mathbf{c} \oplus (\ominus \mathbf{c} \oplus \mathbf{a})\} \ominus \mathrm{gyr}[\mathbf{c}, \ominus \mathbf{c} \oplus \mathbf{a}]\{(\ominus \mathbf{c} \oplus \mathbf{a}) \otimes t\} \\
&\overset{(3)}{=\!=} \mathbf{a} \ominus \mathrm{gyr}[\mathbf{a}, \ominus \mathbf{c}]\{(\ominus \mathbf{c} \oplus \mathbf{a}) \otimes t\} \\
&\overset{(4)}{=\!=} \mathbf{a} \ominus \{\mathrm{gyr}[\mathbf{a}, \ominus \mathbf{c}](\ominus \mathbf{c} \oplus \mathbf{a})\} \otimes t \\
&\overset{(5)}{=\!=} \mathbf{a} \ominus (\mathbf{a} \ominus \mathbf{c}) \otimes t \\
&\overset{(6)}{=\!=} \mathbf{a} \oplus (\ominus \mathbf{a} \oplus \mathbf{c}) \otimes t \,.
\end{aligned}
\tag{3.51}
$$

Derivation of the numbered equalities in (3.35) follows.

(1) This equation follows from the scalar distributive law.

(2) Follows from (1) by the left gyroassociative law.

(3) Follows from (2) by a left cancellation and by (1.130), p. 28.

(4) Follows from (3) by Axiom $(V6)$ of gyrovector spaces.

(5) Follows from (4) by the gyrocommutative law.

(6) Follows from (5) by Theorem 3.4(5), p. 57, and the gyroautomorphic inverse property, Theorem 2.2, p. 33.

□

Lemma 3.20 suggests the following definition.

Definition 3.21.   (Directed Gyrolines). *Let*

$$L = \mathbf{a} \oplus \mathbf{b} \otimes t \tag{3.52}$$

be a gyroline with a parameter $t \in \mathbb{R}$ in a gyrovector space $(G, \oplus, \otimes)$, and let p_1 and p_2 be two distinct points on L,

$$\begin{aligned} p_1 &= \mathbf{a} \oplus \mathbf{b} \otimes t_1 \\ p_2 &= \mathbf{a} \oplus \mathbf{b} \otimes t_2 \,, \end{aligned} \tag{3.53}$$

$\mathbf{a}, \mathbf{b} \in G$, $t_1, t_2 \in \mathbb{R}$. *The gyroline L is directed from p_1 to p_2 if $t_1 < t_2$.*

As an example, the gyroline in (3.49) has the gyroline parameter t and it is directed from $\mathbf{a}$ (where $t = 0$) to $\mathbf{c}$ (where $t = 1$). Similarly, the gyroline in (3.50) has the gyroline parameter $s = 1 - t$, and it is directed from $\mathbf{c}$ (where $s = 0$) to $\mathbf{a}$ (where $s = 1$).

The next lemma relates gyrocollinearity to gyrations.

Lemma 3.22.   *If the three points $\mathbf{a}, \mathbf{b}, \mathbf{c}$ in a gyrovector space $(G, \oplus, \otimes)$ are gyrocollinear then*

$$\mathrm{gyr}[\mathbf{a}, \ominus\mathbf{b}]\mathrm{gyr}[\mathbf{b}, \ominus\mathbf{c}] = \mathrm{gyr}[\mathbf{a}, \ominus\mathbf{c}] \,. \tag{3.54}$$

Proof. By Lemma 3.18,

$$\mathbf{b} = \mathbf{a} \oplus (\ominus\mathbf{a} \oplus \mathbf{c}) \otimes t_0 \tag{3.55}$$

for some $t_0 \in \mathbb{R}$. Hence, by a left cancellation, we have

$$\ominus\mathbf{a} \oplus \mathbf{b} = (\ominus\mathbf{a} \oplus \mathbf{c}) \otimes t_0 \,. \tag{3.56}$$

Hence, by Identity (2.31), p. 38, by the gyroautomorphic inverse property, (2.1), p. 33, and by Axiom (V7) of gyrovector spaces, we have

$$
\begin{aligned}
\mathrm{gyr}[\mathbf{a}, \ominus\mathbf{b}]\mathrm{gyr}[\mathbf{b}, \ominus\mathbf{c}]\mathrm{gyr}[\mathbf{c}, \ominus\mathbf{a}] &= \mathrm{gyr}[\ominus\mathbf{a}\oplus\mathbf{b}, (\mathbf{a}\ominus\mathbf{c})] \\
&= \mathrm{gyr}[\ominus\mathbf{a}\oplus\mathbf{b}, \ominus(\ominus\mathbf{a}\oplus\mathbf{c})] \\
&= \mathrm{gyr}[(\ominus\mathbf{a}\oplus\mathbf{c})\otimes t_0, \ominus(\ominus\mathbf{a}\oplus\mathbf{c})] \\
&= I,
\end{aligned}
\tag{3.57}
$$

from which (3.54) follows by gyration inversion. □

The converse of Lemma 3.22 is not valid, a counterexample being vector spaces. Any vector space is a gyrovector space in which all the gyrations are trivial. Hence, Identity (3.54) holds in vector spaces for any three points $\mathbf{a}$, $\mathbf{b}$, and $\mathbf{c}$ while not every three points of an n-dimensional vector space, $n \geq 2$, are collinear.

The obvious extension of Lemma 3.22 to any number of gyrocollinear points results in the following theorem.

Theorem 3.23.   (The Gyroline Gyration Transitive Law).

Let $\{\mathbf{a}_1, \cdots, \mathbf{a}_n\}$ be a set of n gyrocollinear points in a gyrovector space $(G, \oplus, \otimes)$. Then, we have the telescopic gyration identity

$$
\mathrm{gyr}[\mathbf{a}_1, \ominus\mathbf{a}_2]\mathrm{gyr}[\mathbf{a}_2, \ominus\mathbf{a}_3] \cdots \mathrm{gyr}[\mathbf{a}_{n-1}, \ominus\mathbf{a}_n] = \mathrm{gyr}[\mathbf{a}_1, \ominus\mathbf{a}_n].
\tag{3.58}
$$

Proof. By Lemma 3.22, Identity (3.58) of the theorem holds for $n = 3$. Let us assume, by induction, that Identity (3.58) is valid for some $n = k \geq 3$. Then, Identity (3.58) is valid for $n = k + 1$ as well,

$$
\begin{aligned}
\mathrm{gyr}[\mathbf{a}_1, \ominus\mathbf{a}_2] \ldots \mathrm{gyr}[\mathbf{a}_{k-1}, \ominus\mathbf{a}_k]\mathrm{gyr}[\mathbf{a}_k, \ominus\mathbf{a}_{k+1}] &= \mathrm{gyr}[\mathbf{a}_1, \ominus\mathbf{a}_k]\mathrm{gyr}[\mathbf{a}_k, \ominus\mathbf{a}_{k+1}] \\
&= \mathrm{gyr}[\mathbf{a}_1, \ominus\mathbf{a}_{k+1}].
\end{aligned}
\tag{3.59}
$$

Hence, Identity (3.58) is valid for all $n \geq 3$. □

3.3   GYROMIDPOINTS

The value $t = 1/2$ in Lemma 3.20 gives rise to a special point where the two parameters of the gyroline $\mathbf{b}$, t and $(1 - t)$ coincide. It suggests the following definition.

Definition 3.24.   (Gyromidpoints). *The gyromidpoint $\mathbf{m}_{\mathbf{ac}}$ of any two distinct points $\mathbf{a}$ and $\mathbf{c}$ in a gyrovector space $(G, \oplus, \otimes)$ is given by the equation*

$$
\mathbf{m}_{\mathbf{ac}} = \mathbf{a}\oplus(\ominus\mathbf{a}\oplus\mathbf{c})\otimes\tfrac{1}{2}.
\tag{3.60}
$$

Theorem 3.25. *Let* **a** *and* **c** *be any two points of a gyrovector space* $(G, \oplus, \otimes)$. *Then* $\mathbf{m_{ac}}$ *satisfies the midpoint symmetry condition*

$$\mathbf{m_{ac}} = \mathbf{m_{ca}} , \tag{3.61}$$

as well as the midpoint distance condition

$$\|\mathbf{a} \ominus \mathbf{m_{ac}}\| = \|\mathbf{c} \ominus \mathbf{m_{ac}}\| . \tag{3.62}$$

Proof. By Lemma 3.20, with $t = 1/2$, the two equations

$$\begin{aligned}\mathbf{b} &= \mathbf{a} \oplus (\ominus \mathbf{a} \oplus \mathbf{c}) \otimes \tfrac{1}{2} = \mathbf{m_{ac}}\\ \mathbf{b} &= \mathbf{c} \oplus (\ominus \mathbf{c} \oplus \mathbf{a}) \otimes \tfrac{1}{2} = \mathbf{m_{ca}}\end{aligned} \tag{3.63}$$

are equivalent, thus verifying (3.61).

It follows from (3.63) by left cancellations and the gyrocommutative law, noting the gyroautomorphism property ($V6$) in Def. 3.2, p. 55, that

$$\begin{aligned}\ominus \mathbf{a} \oplus \mathbf{m_{ac}} &= (\ominus \mathbf{a} \oplus \mathbf{c}) \otimes \tfrac{1}{2}\\ \ominus \mathbf{c} \oplus \mathbf{m_{ca}} &= (\ominus \mathbf{c} \oplus \mathbf{a}) \otimes \tfrac{1}{2} = \ominus \mathrm{gyr}[\ominus \mathbf{c}, \mathbf{a}](\ominus \mathbf{a} \oplus \mathbf{c}) \otimes \tfrac{1}{2} ,\end{aligned} \tag{3.64}$$

implying, by Axioms (V1) and (V9) of gyrovector spaces,

$$\begin{aligned}\|\ominus \mathbf{a} \oplus \mathbf{m_{ac}}\| &= \|\ominus \mathbf{a} \oplus \mathbf{c}\| \otimes \tfrac{1}{2}\\ \|\ominus \mathbf{c} \oplus \mathbf{m_{ca}}\| &= \|\ominus \mathbf{a} \oplus \mathbf{c}\| \otimes \tfrac{1}{2} ,\end{aligned} \tag{3.65}$$

thus verifying (3.62). $\qquad\square$

Identities (3.61) and (3.62) justify calling $\mathbf{m_{ac}}$ the gyromidpoint of the points **a** and **c** in Def. 3.24.

Theorem 3.26. *The gyromidpoint of points* **a** *and* **b** *can be written as*

$$\mathbf{m_{ab}} = \tfrac{1}{2} \otimes (\mathbf{a} \boxplus \mathbf{b}) , \tag{3.66}$$

so that

$$\|\mathbf{m_{ab}}\| = \tfrac{1}{2} \otimes \|\mathbf{a} \boxplus \mathbf{b}\| . \tag{3.67}$$

Proof. Identity (3.66) results from the following chain of equations, which are numbered for subsequent derivation:

$$
\begin{aligned}
2\otimes\mathbf{m_{ab}} &\overset{(1)}{=\!=} 2\otimes\{\mathbf{a}\oplus\tfrac{1}{2}\otimes(\ominus\mathbf{a}\oplus\mathbf{b})\} \\
&\overset{(2)}{=\!=} \mathbf{a}\oplus\{(\ominus\mathbf{a}\oplus\mathbf{b})\oplus\mathbf{a}\} \\
&\overset{(3)}{=\!=} \{\mathbf{a}\oplus(\ominus\mathbf{a}\oplus\mathbf{b})\}\oplus\mathrm{gyr}[\mathbf{a},\ominus\mathbf{a}\oplus\mathbf{b}]\mathbf{a} \\
&\overset{(4)}{=\!=} \mathbf{b}\oplus\mathrm{gyr}[\mathbf{a},\ominus\mathbf{a}\oplus\mathbf{b}]\mathbf{a} \\
&\overset{(5)}{=\!=} \mathbf{b}\oplus\mathrm{gyr}[\mathbf{b},\ominus\mathbf{a}]\mathbf{a} \\
&\overset{(6)}{=\!=} \mathbf{b}\boxplus\mathbf{a} \\
&\overset{(7)}{=\!=} \mathbf{a}\boxplus\mathbf{b},
\end{aligned}
\tag{3.68}
$$

implying

$$
\mathbf{m_{ab}} = \tfrac{1}{2}\otimes(\mathbf{a}\boxplus\mathbf{b}) . \tag{3.69}
$$

Derivation of the numbered equalities in (3.68) follows.

(1) Follows from (3.60).

(2) Follows from (1) by the Two-Sum Identity in Theorem 3.9, p. 60.

(3) Follows from (2) by the left gyroassociative law.

(4) Follows from (3) by a left cancellation.

(5) Follows from (4) by Theorem 1.31, p. 28.

(6) Follows from (5) by Def. 1.9, p. 7, of the gyrogroup cooperation $\boxplus$.

(7) Follows from (6) by the commutativity of the cooperation $\boxplus$ according to Theorem 2.3, p. 34.□

3.4 ANALOGIES BETWEEN GYROMIDPOINTS AND MIDPOINTS

The gyromidpoint (3.66) in a gyrovector space $(\mathbb{V}_s, \oplus, \otimes)$ is fully analogous to its Euclidean counterpart, the midpoint $\mathbf{m}^e_{ab}$, in a vector space $(\mathbb{V}, +, \cdot)$,

$$
\mathbf{m}^e_{ab} = \tfrac{1}{2}(\mathbf{a}+\mathbf{b}) . \tag{3.70}
$$

The midpoint $\mathbf{m}^e_{ab}$ is *covariant* with respect to translations in the sense that it satisfies the identity

$$
\mathbf{x}+\mathbf{m}^e_{ab} = \tfrac{1}{2}\{(\mathbf{x}+\mathbf{a})+(\mathbf{x}+\mathbf{b})\} \tag{3.71}
$$

for all $\mathbf{x} \in \mathbb{V}$. Indeed, while Identity (3.71) is immediate, it possesses the important Euclidean geometric interpretation according to which the points $\mathbf{a}$, $\mathbf{b}$ and their midpoint $\mathbf{m}^e_{\mathbf{ab}}$ *vary together* under translations.

In full analogy with (3.71), the gyromidpoint $\mathbf{m}_{\mathbf{ab}}$ in (3.66) satisfies the identity

$$\mathbf{x} \oplus \mathbf{m}_{\mathbf{ab}} = \tfrac{1}{2}\{(\mathbf{x} \oplus \mathbf{a}) \boxplus (\mathbf{x} \oplus \mathbf{b})\} \tag{3.72}$$

for all $\mathbf{x} \in \mathbb{V}_s$. Identity (3.72) possesses the important gyrogeometric interpretation according to which the points $\mathbf{a}$, $\mathbf{b}$ and their gyromidpoint $\mathbf{m}_{\mathbf{ab}}$ *vary together* under left gyrotranslations. Accordingly, the gyromidpoint is said to be *gyrocovariant* with respect to left gyrotranslations. A proof of (3.72) is presented in [61, Sec. 6.6].

Identities (3.69)–(3.72) demonstrate once again that both the gyrogroup operation, $\oplus$, and cooperation, $\boxplus$, are needed in order to capture in gyrogroups and gyrovector spaces analogies with groups and vector spaces.

The point $\mathbf{b} = S^e_{\mathbf{m}^e_{\mathbf{ab}}} \mathbf{a}$ such that $\mathbf{m}^e_{\mathbf{ab}}$ is the midpoint of $\mathbf{b}$ and a given point $\mathbf{a}$ in a vector space $\mathbb{V}$ is said to be the reflection of point $\mathbf{a}$ by the point $\mathbf{m}^e_{\mathbf{ab}}$, and is determined by the equation

$$\mathbf{b} = S^e_{\mathbf{m}^e_{\mathbf{ab}}} \mathbf{a} = 2\mathbf{m}^e_{\mathbf{ab}} - \mathbf{a} \, . \tag{3.73}$$

Indeed, the midpoint of $\mathbf{a}$ and $\mathbf{b} = S^e_{\mathbf{m}^e_{\mathbf{ab}}} \mathbf{a}$ is

$$\tfrac{1}{2}(S^e_{\mathbf{m}^e_{\mathbf{ab}}} \mathbf{a} + \mathbf{a}) = \tfrac{1}{2}\{(2\mathbf{m}^e_{\mathbf{ab}} - \mathbf{a}) + \mathbf{a}\} = \mathbf{m}^e_{\mathbf{ab}} \, . \tag{3.74}$$

In full analogy, the point $\mathbf{b} = S_{\mathbf{m}_{\mathbf{ab}}} \mathbf{a}$ such that $\mathbf{m}_{\mathbf{ab}}$ is the gyromidpoint of $\mathbf{b}$ and a given point $\mathbf{a}$ in a gyrovector space $(\mathbb{V}_s, \oplus, \otimes)$ is said to be the gyroreflection of point $\mathbf{a}$ by the point $\mathbf{m}_{\mathbf{ab}}$, and is determined by the equation

$$\mathbf{b} = S_{\mathbf{m}_{\mathbf{ab}}} \mathbf{a} = 2 \otimes \mathbf{m}_{\mathbf{ab}} \ominus \mathbf{a} \, . \tag{3.75}$$

Indeed, the gyromidpoint of $\mathbf{a}$ and $\mathbf{b} = S_{\mathbf{m}_{\mathbf{ab}}} \mathbf{a}$ is

$$\tfrac{1}{2} \otimes (S_{\mathbf{m}_{\mathbf{ab}}} \mathbf{a} \boxplus \mathbf{a}) = \tfrac{1}{2} \otimes \{(2 \otimes \mathbf{m}_{\mathbf{ab}} \ominus \mathbf{a}) \boxplus \mathbf{a}\} = \mathbf{m}_{\mathbf{ab}} \, , \tag{3.76}$$

as we can see by applying a right cancellation, (1.18), p. 4, followed by the scalar associative law (V4) of gyrovector spaces.

Let $\mathbf{a}$, $\mathbf{b} \in \mathbb{V}$ be any two elements of a vector space $\mathbb{V}$. Then we have the immediate identity

$$S^e_{\mathbf{a}} S^e_{\mathbf{b}} \mathbf{x} = S^e_{S^e_{\mathbf{a}}\mathbf{b}} S^e_{\mathbf{a}} \mathbf{x} \tag{3.77}$$

for all $\mathbf{x} \in \mathbb{V}$.

Remarkably, property (3.77) of midpoints in a vector space remains valid for gyromidpoints in a gyrovector space, as we see from the following theorem.

Theorem 3.27. *Let $\mathbf{p}$ and $\mathbf{v}$ be any two elements of a gyrovector space $(\mathbb{V}_s, \oplus, \otimes)$, and let $S_{\mathbf{p}} : \mathbb{V}_s \to \mathbb{V}_s$ be a self map of $\mathbb{V}_s$ given by*

$$S_{\mathbf{p}} \mathbf{v} = 2 \otimes \mathbf{p} \ominus \mathbf{v} \, . \tag{3.78}$$

Then,

$$S_{S_\mathbf{a}\mathbf{b}}S_\mathbf{a}\mathbf{x} = S_\mathbf{a}S_\mathbf{b}\mathbf{x} \tag{3.79}$$

for all $\mathbf{a}, \mathbf{b}, \mathbf{x} \in \mathbb{V}_s$.

Proof. On the one hand, we have the chain of equations

$$
\begin{aligned}
S_{S_\mathbf{a}\mathbf{b}}S_\mathbf{a}\mathbf{x} &\overset{(1)}{=\!=\!=} S_{S_\mathbf{a}\mathbf{b}}(2\otimes\mathbf{a}\ominus\mathbf{x}) \\
&\overset{(2)}{=\!=\!=} 2\otimes S_\mathbf{a}\mathbf{b}\ominus(2\otimes\mathbf{a}\ominus\mathbf{x}) \\
&\overset{(3)}{=\!=\!=} 2\otimes(2\otimes\mathbf{a}\ominus\mathbf{b})\ominus(2\otimes\mathbf{a}\ominus\mathbf{x}) \\
&\overset{(4)}{=\!=\!=} (2\otimes\mathbf{a}\oplus\{\ominus2\otimes\mathbf{b}\oplus2\otimes\mathbf{a}\})\ominus(2\otimes\mathbf{a}\ominus\mathbf{x}) \\
&\overset{(5)}{=\!=\!=} \mathrm{gyr}[2\otimes\mathbf{a}, \ominus2\otimes\mathbf{b}\oplus2\otimes\mathbf{a}]\{(\ominus2\otimes\mathbf{b}\oplus2\otimes\mathbf{a})\oplus\mathbf{x}\} \\
&\overset{(6)}{=\!=\!=} \mathrm{gyr}[2\otimes\mathbf{a}, \ominus2\otimes\mathbf{b}]\{(\ominus2\otimes\mathbf{b}\oplus2\otimes\mathbf{a})\oplus\mathbf{x}\} \\
&\overset{(7)}{=\!=\!=} (2\otimes\mathbf{a}\ominus2\otimes\mathbf{b})\oplus\mathrm{gyr}[2\otimes\mathbf{a}, \ominus2\otimes\mathbf{b}]\mathbf{x}
\end{aligned}
\tag{3.80}
$$

for all $\mathbf{a}, \mathbf{b}, \mathbf{x} \in \mathbb{V}_s$. Derivation of the numbered equalities in (3.80) follows.

(1) Follows by the application of the map $S_\mathbf{a}$.

(2) Follows from (1) by the application of the map $S_{S_\mathbf{a}\mathbf{b}}$.

(3) Follows from (2) by the application of the map $S_\mathbf{a}$.

(4) Follows from (3) by the Two-Sum Identity in Theorem 3.9, p. 60.

(5) Follows from (4) by the Gyrotranslation Theorem 2.11, p. 37.

(6) Follows from (5) by the right loop property. Indeed, the application of the right loop property to the gyration in (6) gives the gyration in (5).

(7) Follows from (6) by applying the gyration in (6) term-wise to each of the two terms in {...}, followed by an application of the gyrocommutative law.

On the other hand, we have the chain of equation

$$
\begin{aligned}
S_\mathbf{a}S_\mathbf{b}\mathbf{x} &\overset{(1)}{=\!=\!=} S_\mathbf{a}(2\otimes\mathbf{b}\ominus\mathbf{x}) \\
&\overset{(2)}{=\!=\!=} 2\otimes\mathbf{a}\ominus(2\otimes\mathbf{b}\ominus\mathbf{x}) \\
&\overset{(3)}{=\!=\!=} 2\otimes\mathbf{a}\oplus(\ominus2\otimes\mathbf{b}\oplus\mathbf{x}) \\
&\overset{(4)}{=\!=\!=} (2\otimes\mathbf{a}\ominus2\otimes\mathbf{b})\oplus\mathrm{gyr}[2\otimes\mathbf{a}, \ominus2\otimes\mathbf{b}]\mathbf{x}
\end{aligned}
\tag{3.81}
$$

for all $\mathbf{a}, \mathbf{b}, \mathbf{x} \in \mathbb{V}_s$. Derivation of the numbered equalities in (3.81) follows.

(1) Follows from the application of the map $S_{\mathbf{b}}$.

(2) Follows from (1) by the application of the map $S_{\mathbf{a}}$.

(3) Follows from (2) by the gyroautomorphic inverse property, Theorem 2.2, p. 33.

(4) Follows from (3) by the left gyroassociative law.

Finally, comparing the extreme right-hand sides of (3.80) and (3.81) we obtain the result (3.79) of the theorem.   □

3.5   GYROGEODESICS

In the presence of metric, geodesics are the length minimizing curves. In full analogy, in the presence of gyrometric, gyrogeodesics are the gyrolength minimizing curves.

Accordingly, following the study of Möbius gyrovector spaces in Sec. 3.6, we will find in Sec. 3.7 that Möbius gyrolines are gyrogeodesics that coincide with the well-known geodesics of the Poincaré ball model of hyperbolic geometry. Similarly, following the study of Einstein gyrovector spaces in Sec. 3.8 we will find in Sec. 3.9 that Einstein gyrolines are gyrogeodesics that coincide with the well-known geodesics of the Beltrami-Klein ball model of hyperbolic geometry.

The following theorem gives a condition that reduces the gyrotriangle inequality (3.23) to an equality, enabling us to recognize gyrolines as gyrogeodesics.

Theorem 3.28.   (The Gyrotriangle Equality). *If a point* $\mathbf{b}$ *lies between two points* $\mathbf{a}$ *and* $\mathbf{c}$ *in a gyrovector space* $(G, \oplus, \otimes)$ *then*

$$\|\ominus\mathbf{a}\oplus\mathbf{c}\| = \|\ominus\mathbf{a}\oplus\mathbf{b}\|\oplus\|\ominus\mathbf{b}\oplus\mathbf{c}\|\,. \tag{3.82}$$

Proof. If $\mathbf{b}$ lies between $\mathbf{a}$ and $\mathbf{c}$ then, by Lemma 3.19, p. 66,

$$\mathbf{b} = \mathbf{a}\oplus(\ominus\mathbf{a}\oplus\mathbf{c})\otimes t_0 \tag{3.83}$$

for some $0 < t_0 < 1$, and hence, by Lemma 3.20

$$\mathbf{b} = \mathbf{c}\oplus(\ominus\mathbf{c}\oplus\mathbf{a})\otimes(1 - t_0)\,. \tag{3.84}$$

Hence, by left cancellations, (3.83) – (3.84) give

$$\ominus\mathbf{a}\oplus\mathbf{b} = (\ominus\mathbf{a}\oplus\mathbf{c})\otimes t_0$$
$$\ominus\mathbf{c}\oplus\mathbf{b} = (\ominus\mathbf{c}\oplus\mathbf{a})\otimes(1 - t_0)\,. \tag{3.85}$$

Taking magnitudes, noting the homogeneity property ($V9$) of gyrovector spaces in Def. 3.2, p. 55, (3.85) gives

$$\|\ominus\mathbf{a}\oplus\mathbf{b}\| = \|\ominus\mathbf{a}\oplus\mathbf{c}\|\otimes t_0$$
$$\|\ominus\mathbf{b}\oplus\mathbf{c}\| = \|\ominus\mathbf{a}\oplus\mathbf{c}\|\otimes(1-t_0)\,,$$

(3.86)

so that, by the distributive law ($V8$) of real vector spaces in Def. 3.2(3), p. 55,

$$\|\ominus\mathbf{a}\oplus\mathbf{b}\|\oplus\|\ominus\mathbf{b}\oplus\mathbf{c}\| = \|\ominus\mathbf{a}\oplus\mathbf{c}\|\otimes t_0\oplus\|\ominus\mathbf{a}\oplus\mathbf{c}\|\otimes(1-t_0)$$
$$= \|\ominus\mathbf{a}\oplus\mathbf{c}\|\otimes\{t_0 + (1-t_0)\}$$
$$= \|\ominus\mathbf{a}\oplus\mathbf{c}\|\,.$$

(3.87)

$\square$

Remark 3.29. Comparing Theorem 3.28 with Theorem 3.11, p. 61, we see that points $\mathbf{b}$ between two given points $\mathbf{a}$ and $\mathbf{c}$ in a gyrovector space $(G, \oplus, \otimes)$ (i) turn the gyrotriangle inequality into an equality, and hence (ii) minimize the gyrodistance gyrosum $\|\ominus\mathbf{a}\oplus\mathbf{b}\|\oplus\|\ominus\mathbf{b}\oplus\mathbf{c}\|$. Since these points $\mathbf{b}$ lie between $\mathbf{a}$ and $\mathbf{c}$ on the gyroline generated by $\mathbf{a}$ and $\mathbf{c}$, gyrolines are curves that minimize gyrodistance gyrosum. Hence, gyrolines are also called gyrogeodesics.

In Sec. 3.6 we will realize the abstract gyrovector space by Möbius gyrovector spaces, and in Sec. 3.7 we will find that Möbius gyrolines, which are Möbius gyrogeodesics, turn out to be the well-known geodesics of the Poincaré ball model of hyperbolic geometry.

Similarly, in Sec. 3.8 we will realize the abstract gyrovector space by Einstein gyrovector spaces, and in Sec. 3.9 we will find that Einstein gyrolines, which are Einstein gyrogeodesics, turn out to be the well-known geodesics of the Beltrami-Klein ball model of hyperbolic geometry.

3.6   MÖBIUS GYROVECTOR SPACES

We show in this section that Möbius gyrogroups $(\mathbb{V}_s, \oplus_{_M})$ admit scalar multiplication $\otimes_{_M}$, giving rise to Möbius gyrovector spaces $(\mathbb{V}_s, \oplus_{_M}, \otimes_{_M})$. The latter, in turn, form the algebraic setting for the Poincaré ball model of hyperbolic geometry just as vector spaces form the algebraic setting for the standard model of Euclidean geometry as we indicate in Sec. 3.7.

Definition 3.30.   (Möbius Scalar Multiplication). *Let $(\mathbb{V}_s, \oplus_{_M})$ be a Möbius gyrogroup. The Möbius scalar multiplication $r\otimes_{_M}\mathbf{v} = \mathbf{v}\otimes_{_M}r$ in $\mathbb{V}_s$ is given by the equation*

$$r\otimes_{_M}\mathbf{v} = s\frac{\left(1+\dfrac{\|\mathbf{v}\|}{s}\right)^r - \left(1-\dfrac{\|\mathbf{v}\|}{s}\right)^r}{\left(1+\dfrac{\|\mathbf{v}\|}{s}\right)^r + \left(1-\dfrac{\|\mathbf{v}\|}{s}\right)^r}\frac{\mathbf{v}}{\|\mathbf{v}\|}$$

(3.88)

$$= s\tanh(r\,\tanh^{-1}\frac{\|\mathbf{v}\|}{s})\frac{\mathbf{v}}{\|\mathbf{v}\|}\,,$$

where $r \in \mathbb{R}$, $\mathbf{v} \in \mathbb{V}_s$, $\mathbf{v} \neq \mathbf{0}$, and where $r \otimes_M \mathbf{0} = \mathbf{0}$.

Möbius scalar multiplication in Def. 3.30 is obtained by calculating $n \otimes \mathbf{v} := \mathbf{v} \oplus \ldots \oplus \mathbf{v}$ (n terms) and formally replacing $n \in \mathbb{N}$ by $r \in \mathbb{R}$.

The following theorem demonstrates that the resulting scalar multiplication indeed gives the desired gyrovector space.

Theorem 3.31. *A Möbius gyrogroup $(\mathbb{V}_s, \oplus_M)$ with Möbius scalar multiplication $\otimes_M$ in $\mathbb{V}_s$ forms a gyrovector space $(\mathbb{V}_s, \oplus_M, \otimes_M)$.*

We will verify the gyrovector space axioms in Def. 3.2, p. 55, for any Möbius gyrovector space $(\mathbb{V}_s, \oplus_M, \otimes_M)$.

(V1) Inner Product Gyroinvariance:

Axiom $(V1)$ is satisfied as verified in (1.32), p. 10.

(V2) Identity Scalar Multiplication:

Let $\mathbf{a} \in \mathbb{V}_s$. If $\mathbf{a} = \mathbf{0}$ then $1 \otimes_M \mathbf{0} = \mathbf{0}$ by definition, so that Axiom $(V2)$ is satisfied. For $\mathbf{a} \neq \mathbf{0}$ we have

$$1 \otimes_M \mathbf{a} = s \tanh(\tanh^{-1} \frac{\|\mathbf{a}\|}{s}) \frac{\mathbf{a}}{\|\mathbf{a}\|}$$
$$= \mathbf{a}\,,$$

(3.89)

thus verifying Axiom $(V2)$ of gyrovector spaces.

(V3) Scalar Distributive Law:

Let $r_1, r_2 \in \mathbb{R}$. If $\mathbf{a} = \mathbf{0}$ then $(r_1 + r_2) \otimes_M \mathbf{a} = \mathbf{0}$ and $r_1 \otimes_M \mathbf{a} \oplus_M r_2 \otimes_M \mathbf{a} = \mathbf{0} \oplus_M \mathbf{0} = \mathbf{0}$ so that Axiom $(V3)$ is satisfied. For $\mathbf{a} \neq \mathbf{0}$ we have, by (3.88), the addition formula of the hyperbolic tangent function, and (2.54), p. 42,

$$(r_1 + r_2) \otimes_M \mathbf{a} = s \tanh([r_1 + r_2] \tanh^{-1} \frac{\|\mathbf{a}\|}{s}) \frac{\mathbf{a}}{\|\mathbf{a}\|}$$

$$= s \tanh(r_1 \tanh^{-1} \frac{\|\mathbf{a}\|}{s} + r_2 \tanh^{-1} \frac{\|\mathbf{a}\|}{s}) \frac{\mathbf{a}}{\|\mathbf{a}\|}$$

$$= s \frac{\tanh(r_1 \tanh^{-1} \frac{\|\mathbf{a}\|}{s}) + \tanh(r_2 \tanh^{-1} \frac{\|\mathbf{a}\|}{s})}{1 + \tanh(r_1 \tanh^{-1} \frac{\|\mathbf{a}\|}{s}) \tanh(r_2 \tanh^{-1} \frac{\|\mathbf{a}\|}{s})} \frac{\mathbf{a}}{\|\mathbf{a}\|}$$

$$= \frac{r_1 \otimes_M \mathbf{a} + r_2 \otimes_M \mathbf{a}}{1 + \frac{1}{s^2} \|r_1 \otimes_M \mathbf{a}\| \|r_2 \otimes_M \mathbf{a}\|} = r_1 \otimes_M \mathbf{a} \oplus_M r_2 \otimes_M \mathbf{a}\,,$$

(3.90)

thus verifying Axiom $(V3)$ of gyrovector spaces.

(V4) Scalar Associative Law:

For $\mathbf{a} = \mathbf{0}$ the validity of Axiom (V4) of gyrovector spaces is obvious. Assuming $\mathbf{a} \neq \mathbf{0}$, let us use the notation

$$\mathbf{b} = r_2 \otimes_{\mathrm{M}} \mathbf{a} = s\ \tanh(r_2 \tanh^{-1} \frac{\|\mathbf{a}\|}{s}) \frac{\mathbf{a}}{\|\mathbf{a}\|}\ , \tag{3.91}$$

so that

$$\frac{\|\mathbf{b}\|}{s} = \tanh(r_2 \tanh^{-1} \frac{\|\mathbf{a}\|}{s}) \tag{3.92}$$

and

$$\frac{\mathbf{b}}{\|\mathbf{b}\|} = \frac{\mathbf{a}}{\|\mathbf{a}\|}\ . \tag{3.93}$$

Then

$$
\begin{aligned}
r_1 \otimes_{\mathrm{M}} (r_2 \otimes_{\mathrm{M}} \mathbf{a}) &= r_1 \otimes_{\mathrm{M}} \mathbf{b} \\[2mm]
&= s\ \tanh(r_1 \tanh^{-1} \frac{\|\mathbf{b}\|}{s}) \frac{\mathbf{b}}{\|\mathbf{b}\|} \\[2mm]
&= s\ \tanh(r_1 \tanh^{-1}(\tanh(r_2 \tanh^{-1} \frac{\|\mathbf{a}\|}{s}))) \frac{\mathbf{a}}{\|\mathbf{a}\|} \\[2mm]
&= s\ \tanh(r_1 r_2 \tanh^{-1} \frac{\|\mathbf{a}\|}{s}) \frac{\mathbf{a}}{\|\mathbf{a}\|} \\[2mm]
&= (r_1 r_2) \otimes_{\mathrm{M}} \mathbf{a}\ .
\end{aligned}
\tag{3.94}
$$

(V5) Scaling Property:

It follows from the Möbius scalar multiplication definition that

$$
\begin{aligned}
|r| \otimes_{\mathrm{M}} \mathbf{a} &= s\ \tanh(|r| \tanh^{-1} \frac{\|\mathbf{a}\|}{s}) \frac{\mathbf{a}}{\|\mathbf{a}\|} \\[2mm]
|r| \otimes_{\mathrm{M}} \|\mathbf{a}\| &= s\ \tanh(|r| \tanh^{-1} \frac{\|\mathbf{a}\|}{s}) \\[2mm]
\|r \otimes_{\mathrm{M}} \mathbf{a}\| &= |s\ \tanh(r \tanh^{-1} \frac{\|\mathbf{a}\|}{s})| = s\ \tanh(|r| \tanh^{-1} \frac{\|\mathbf{a}\|}{s})
\end{aligned}
\tag{3.95}
$$

for $r \in \mathbb{R}$ and $\mathbf{a} \neq \mathbf{0}$. Axiom (V5) follows immediately from the first and the third identity in (3.95).

(V6) Gyroautomorphism Property:

If $\mathbf{v} = \mathbf{0}$ then Axiom (V6) clearly holds. Let $\mathbb{V}$ be the carrier vector space of the ball $\mathbb{V}_s$, and let $\phi : \mathbb{V} \rightarrow \mathbb{V}$ be an invertible linear map that keeps invariant the inner product and, hence, the norm in $\mathbb{V}$. Furthermore, let $\mathbf{v} \in \mathbb{V}_s \subset \mathbb{V}, \mathbf{v} \neq \mathbf{0}$.

Then, by (3.88),

$$\phi(r\otimes_M \mathbf{v}) = \phi\left(s\frac{\left(1+\frac{\|\mathbf{v}\|}{s}\right)^r - \left(1-\frac{\|\mathbf{v}\|}{s}\right)^r}{\left(1+\frac{\|\mathbf{v}\|}{s}\right)^r + \left(1-\frac{\|\mathbf{v}\|}{s}\right)^r}\frac{\mathbf{v}}{\|\mathbf{v}\|}\right)$$

$$= s\frac{\left(1+\frac{\|\mathbf{v}\|}{s}\right)^r - \left(1-\frac{\|\mathbf{v}\|}{s}\right)^r}{\left(1+\frac{\|\mathbf{v}\|}{s}\right)^r + \left(1-\frac{\|\mathbf{v}\|}{s}\right)^r}\frac{\phi(\mathbf{v})}{\|\mathbf{v}\|} \qquad (3.96)$$

$$= s\frac{\left(1+\frac{\|\phi(\mathbf{v})\|}{s}\right)^r - \left(1-\frac{\|\phi(\mathbf{v})\|}{s}\right)^r}{\left(1+\frac{\|\phi(\mathbf{v})\|}{s}\right)^r + \left(1-\frac{\|\phi(\mathbf{v})\|}{s}\right)^r}\frac{\phi(\mathbf{v})}{\|\phi(\mathbf{v})\|}$$

$$= r\otimes_M \phi(\mathbf{v}),$$

$r \in \mathbb{R}$, $\mathbf{v} \in \mathbb{V}_s$. It should be noted that while ϕ is a self map of $\mathbb{V}$, it is restricted in (3.96) to the ball $\mathbb{V}_s \subset \mathbb{V}$.

Möbius gyrations of the ball $\mathbb{V}_s \subset \mathbb{V}$ can be extended to invertible linear maps of $\mathbb{V}$ onto itself, as remarked in Remark 1.12, p. 9. Hence, the map ϕ in (3.96) can be realized by any Möbius gyration of the ball $\mathbb{V}_s \subset \mathbb{V}$. In the special case when the map ϕ of $\mathbb{V}_s$ is a gyration of $\mathbb{V}_s$, $\phi(\mathbf{a}) = \text{gyr}[\mathbf{u}, \mathbf{v}]\mathbf{a}$ for $\mathbf{a} \in \mathbb{V}_s$ and arbitrary, but fixed, $\mathbf{u}, \mathbf{v} \in \mathbb{V}_s$, Identity (3.96) reduces to Axiom $(V6)$, thus verifying its validity.

(V7) Identity Gyroautomorphism:

If either $r_1 = 0$ or $r_2 = 0$ or $\mathbf{v} = \mathbf{0}$ then Axiom $(V7)$ holds by Theorem 1.13(13), p. 11. Assuming $r_1 \neq 0$, $r_2 \neq 0$, and $\mathbf{v} \neq \mathbf{0}$, the vectors $r_1\otimes_M \mathbf{a}$ and $r_2\otimes_M \mathbf{a}$ are parallel in $\mathbb{V}_s \subset \mathbb{V}$. Hence, the gyration $\text{gyr}[r_1\otimes_M \mathbf{v}, r_2\otimes_M \mathbf{v}]$ is trivial, by (1.27)–(1.28), p. 9, thus verifying Axiom $(V7)$.

(V8) Vector Space:

The vector space $(\|G\|, \oplus_M, \otimes_M)$ is studied in [5].

(V9) Homogeneity Property:

Axiom $(V9)$ follows immediately from the second and the third identities in (3.95).

(V10) Gyrotriangle Inequality:

Axiom $(V10)$ is Theorem 2.17, p. 43. □

As an example of Möbius scalar multiplication we present the Möbius half,

$$\tfrac{1}{2}\otimes_{\mathrm{M}}\mathbf{v} = \frac{\gamma_{\mathbf{v}}}{1+\gamma_{\mathbf{v}}}\mathbf{v}\,. \tag{3.97}$$

It satisfies the identity

$$\gamma_{(1/2)\otimes\mathbf{v}} = \sqrt{\frac{1+\gamma_{\mathbf{v}}}{2}}\,, \tag{3.98}$$

where $\gamma_{\mathbf{v}}$ is the gamma factor (2.73), p. 46. In accordance with the scalar associative law of gyrovector spaces, we have

$$2\otimes_{\mathrm{M}}(\tfrac{1}{2}\otimes_{\mathrm{M}}\mathbf{v}) = 2\otimes_{\mathrm{M}}\frac{\gamma_{\mathbf{v}}}{1+\gamma_{\mathbf{v}}}\mathbf{v} = \frac{\gamma_{\mathbf{v}}}{1+\gamma_{\mathbf{v}}}\mathbf{v}\oplus_{\mathrm{M}}\frac{\gamma_{\mathbf{v}}}{1+\gamma_{\mathbf{v}}}\mathbf{v} = \mathbf{v}\,. \tag{3.99}$$

3.7   MÖBIUS GYROLINES

The unique Möbius gyroline L_{AB}, Fig. 3.1, that passes through two given points A and B in a Möbius gyrovector space $\mathbb{V}_s = (\mathbb{V}_s, \oplus_{\mathrm{M}}, \otimes_{\mathrm{M}})$ is represented by the equation

$$L_{AB} = A\oplus_{\mathrm{M}}(\ominus_{\mathrm{M}}A\oplus_{\mathrm{M}}B)\otimes_{\mathrm{M}}t\,, \tag{3.100}$$

$t \in \mathbb{R}$. It is a map of the real line $\mathbb{R}$ into the gyrovector space $\mathbb{V}_s$. Gyrolines in a Möbius gyrovector space $(\mathbb{R}^n_s, \oplus_{\mathrm{M}}, \otimes_{\mathrm{M}})$ turn out to be the well-known geodesics of the Poincaré ball model of hyperbolic geometry. They are Euclidean circular arcs that approach the boundary of the disc orthogonally, as shown in Figs. 3.1–3.2.

Figure 3.2 presents the gyrosegment AB that joins the points A and B in the Möbius gyrovector plane $(\mathbb{R}^2_{s=1}, \oplus_{\mathrm{M}}, \otimes_{\mathrm{M}})$ along with its gyromidpoint M_{AB}, and a generic point P lying between A and B. Since the points A, B, P are gyrocollinear, they satisfy the gyrotriangle equality, Theorem 3.28, shown in the figure.

Considering the parameter $t\in\mathbb{R}$ of (3.100) as "time", the generic point P of the gyroline L_{AB} in Fig. 3.2 travels along the gyrosegment AB during the time interval $0 \le t \le 1$. It reaches the point A at "time" $t = 0$, it reaches the gyromidpoint M_{AB} at "time" $t = 1/2$, as we see from (3.60), p. 69, and owing to the left cancellation law, it reaches the point B at "time" $t = 1$.

The gyroline equation (3.100) is coordinate free. However, in order to draw graphs of gyrolines, as in Figs. 3.1–3.2, we must introduce a coordinate system into $\mathbb{R}^n_s$. Accordingly, we introduce into the unit discs in Figs. 3.1–3.2 rectangular Cartesian x_1x_2-coordinates with the condition $x_1^2 + x_2^2 < s^2$ ($s = 1$ in the figures). Relative to the x_1, x_2-coordinates we select, as an example, the points $A = (-0.5, -0.1)$ and $B = (0, -0.5)$ in the unit disc of Fig. 3.1, and the points $A = (-0.5, 0)$ and $B = (0.1, 0.8)$ in the unit disc of Fig. 3.2. We, then, employ Möbius addition $\oplus_{\mathrm{M}}$ and Möbius scalar multiplication $\otimes_{\mathrm{M}}$ for various values of $t\in\mathbb{R}$ in order to calculate points of the gyroline graphs in Figs. 3.1–3.2.

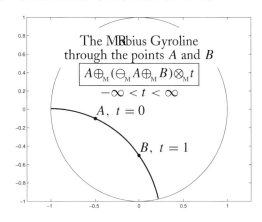

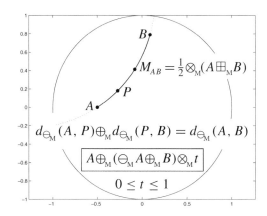

Figure 3.1: The unique gyroline L_{AB} in a Möbius gyrovector space $(\mathbb{R}_s^n, \oplus_M, \otimes_M)$ through two given points A and B. The case of the Möbius gyrovector plane, when $\mathbb{R}_s^n = \mathbb{R}_{s=1}^2$ is the real open unit disc, is shown.

Figure 3.2: The gyroline in $(\mathbb{R}_s^n, \oplus_M, \otimes_M)$ through the points A and B. The point P is a generic point on the gyrosegment AB, and M_{AB} is its gyromidpoint. P lies between A and B, satisfying the gyrotriangle equality, (3.82).

An important contact with the elementary differential geometry of the Poincaré disc model of hyperbolic geometry is provided by the so called *Riemannian gyroline element* of Möbius gyrometric (3.21), p. 61, of the Möbius gyrovector plane $(\mathbb{R}_c^2, \oplus_M, \otimes_M)$. Let $\mathbf{v} \in \mathbb{R}_c^2$ be an element of the disc which, relative to the Cartesian x_1, x_2-coordinates of the disc, is written as the column vector $(x_1, x_2)^t$, where exponent t denotes transposition. Accordingly, the differential $d\mathbf{v}$ of $\mathbf{v}$ is written as $d\mathbf{v} = (dx_1, dx_2)^t$. We now consider the *gyrodifferential*

$$d\mathbf{s} = (\mathbf{v} + d\mathbf{v}) \ominus_M \mathbf{v} = \begin{pmatrix} x_1 + dx_1 \\ x_2 + dx_2 \end{pmatrix} \ominus_M \begin{pmatrix} x_1 \\ x_2 \end{pmatrix}, \tag{3.101}$$

the square of which, $d\mathbf{s}^2 = d\mathbf{s} \cdot d\mathbf{s} = \|d\mathbf{s}\|^2$, takes the form

$$d\mathbf{s}^2 = \|(\mathbf{v} + d\mathbf{v}) \ominus_M \mathbf{v}\|^2 = \left\| \begin{pmatrix} x_1 + dx_1 \\ x_2 + dx_2 \end{pmatrix} \ominus_M \begin{pmatrix} x_1 \\ x_2 \end{pmatrix} \right\|^2, \tag{3.102}$$

and has an important gyrogeometric interpretation. Indeed, $d\mathbf{s}^2$ of (3.102) is the squared gyrodistance between the two infinitesimally nearby points $\mathbf{v}$ and $\mathbf{v} + d\mathbf{v}$ in a Möbius gyrovector plane, called a *Riemannian gyroline element*. Remarkably, an explicit calculation of $d\mathbf{s}^2$ in (3.102) reveals that $d\mathbf{s}^2$ is the quadratical differential form

$$d\mathbf{s}^2 = \frac{dx_1^2 + dx_2^2}{\left(1 - \dfrac{r^2}{c^2}\right)^2}, \tag{3.103}$$

where $r^2 = x_1^2 + x_2^2$.

What is remarkable in the Riemannian gyroline element (3.103) of Möbius gyrometric $d(\mathbf{u}, \mathbf{v}) = \|\mathbf{u} \ominus_{_M} \mathbf{v}\|$ is that it turns out to be identical with the well-known Riemannian line element of the hyperbolic metric of the Poincaré disc model of hyperbolic geometry. The special case of (3.103) when $c = 2$ is presented in detail in [30, p. 226]. Furthermore, (3.103) along with its natural generalization to higher dimensions is studied in detail in [61, Sec. 7.3].

3.8   EINSTEIN GYROVECTOR SPACES

We show in this section that Einstein gyrogroups $(\mathbb{V}_s, \oplus_{_E})$ admit scalar multiplication $\otimes_{_E}$, giving rise to Einstein gyrovector spaces $(\mathbb{V}_s, \oplus_{_E}, \otimes_{_E})$. The latter, in turn, form the algebraic setting for the Beltrami-Klein ball model of hyperbolic geometry just as vector spaces form the algebraic setting for the standard model of Euclidean geometry as we indicate in Sec. 3.9.

Definition 3.32.   (Einstein Scalar Multiplication). *Let $(\mathbb{V}_s, \oplus_{_E})$ be an Einstein gyrogroup. The Einstein scalar multiplication $r \otimes_{_E} \mathbf{v} = \mathbf{v} \otimes_{_E} r$ in $\mathbb{V}_s$ is given by the equation*

$$r \otimes_{_E} \mathbf{v} = s \frac{(1 + \|\mathbf{v}\|/s)^r - (1 - \|\mathbf{v}\|/s)^r}{(1 + \|\mathbf{v}\|/s)^r + (1 - \|\mathbf{v}\|/s)^r} \frac{\mathbf{v}}{\|\mathbf{v}\|} = s \tanh(r \tanh^{-1} \frac{\|\mathbf{v}\|}{s}) \frac{\mathbf{v}}{\|\mathbf{v}\|}, \qquad (3.104)$$

where $r \in \mathbb{R}$, $\mathbf{v} \in \mathbb{V}_s$, $\mathbf{v} \neq \mathbf{0}$; and $r \otimes_{_E} \mathbf{0} = \mathbf{0}$.

Interestingly, the scalar multiplication that Möbius and Einstein addition admit coincide. This coincidence stems from the fact that for parallel vectors in the ball, Möbius addition and Einstein addition coincide as well, as we see from (2.54), p. 42, and (2.76), p. 46.

Owing to Identity (2.46), p. 41, Einstein scalar multiplication can also be written in terms of the gamma factor (2.45), p. 41, as

$$r \otimes_{_E} \mathbf{v} = \frac{1 - (\gamma_{\mathbf{v}} - \sqrt{\gamma_{\mathbf{v}}^2 - 1})^{2r}}{1 + (\gamma_{\mathbf{v}} - \sqrt{\gamma_{\mathbf{v}}^2 - 1})^{2r}} \frac{\gamma_{\mathbf{v}}}{\sqrt{\gamma_{\mathbf{v}}^2 - 1}} \mathbf{v}, \qquad (3.105)$$

$\mathbf{v} \neq \mathbf{0}$.

As an example, (3.105) specializes to Einstein half when $r = 1/2$, obtaining

$$\tfrac{1}{2} \otimes_{_E} \mathbf{v} = \frac{\gamma_{\mathbf{v}}}{1 + \gamma_{\mathbf{v}}} \mathbf{v}. \qquad (3.106)$$

The gamma factor of $r \otimes_{_E} \mathbf{v}$ is expressible in terms of the gamma factor of $\mathbf{v}$ by the identity

$$\gamma_{r \otimes_{_E} \mathbf{v}} = \frac{1}{2} \gamma_{\mathbf{v}}^r \left\{ \left(1 + \frac{\|\mathbf{v}\|}{s}\right)^r + \left(1 - \frac{\|\mathbf{v}\|}{s}\right)^r \right\} \qquad (3.107)$$

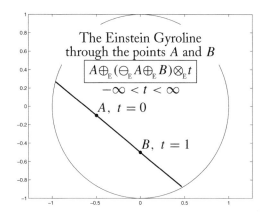

Figure 3.3: The unique gyroline L_{AB} in an Einstein gyrovector space $(\mathbb{R}^n_s, \oplus_{\mathrm{E}}, \otimes_{\mathrm{E}})$ through two given points A and B. The case of the Einstein gyrovector plane, when $\mathbb{R}^n_s = \mathbb{R}^2_{s=1}$ is the real open unit disc, is shown.

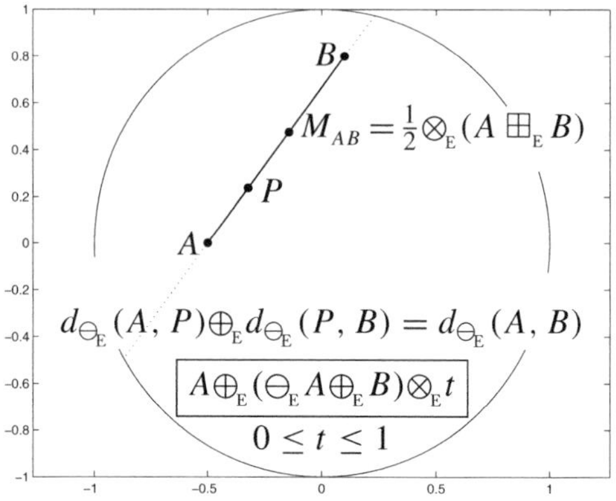

Figure 3.4: The gyroline in $(\mathbb{R}^n_s, \oplus_{\mathrm{E}}, \otimes_{\mathrm{E}})$ through the points A and B. The point P is a generic point on the gyrosegment AB, and M_{AB} is its gyromidpoint. P lies between A and B, satisfying the gyrotriangle equality (3.82).

and hence, by (3.104),

$$\gamma_{r\otimes_{\mathrm{E}}\mathbf{v}}(r\otimes_{\mathrm{E}}\mathbf{v}) = \frac{1}{2}\gamma^r_{\mathbf{v}}\left\{\left(1 + \frac{\|\mathbf{v}\|}{s}\right)^r - \left(1 - \frac{\|\mathbf{v}\|}{s}\right)^r\right\}\frac{\mathbf{v}}{\|\mathbf{v}\|} \tag{3.108}$$

for $\mathbf{v} \neq \mathbf{0}$. The special case of $r = 2$ is of particular interest,

$$\gamma_{2\otimes_{\mathrm{E}}\mathbf{v}}(2\otimes_{\mathrm{E}}\mathbf{v}) = 2\gamma^2_{\mathbf{v}}\mathbf{v}. \tag{3.109}$$

Noting (2.46), p. 41, we have from (3.104),

$$2\otimes_{\mathrm{E}}\mathbf{v} = \frac{2\gamma^2_{\mathbf{v}}}{2\gamma^2_{\mathbf{v}} - 1}\mathbf{v}, \tag{3.110}$$

so that

$$\gamma_{2\otimes_{\mathrm{E}}\mathbf{v}} = 2\gamma^2_{\mathbf{v}} - 1 = \frac{1 + \|\mathbf{v}\|^2/s^2}{1 - \|\mathbf{v}\|^2/s^2}. \tag{3.111}$$

3.9   EINSTEIN GYROLINES

The unique Einstein gyroline L_{AB}, Fig. 3.3, that passes through two given points A and B in an Einstein gyrovector space $\mathbb{V}_s = (\mathbb{V}_s, \oplus_{\mathrm{E}}, \otimes_{\mathrm{E}})$ is represented by the equation

$$L_{AB} = A\oplus_{\mathrm{E}}(\ominus_{\mathrm{E}}A\oplus_{\mathrm{E}}B)\otimes_{\mathrm{E}}t, \tag{3.112}$$

$t \in \mathbb{R}$. It is a map of the real line $\mathbb{R}$ into the gyrovector space $\mathbb{V}_s$. Gyrolines in an Einstein gyrovector space $(\mathbb{R}^n_s, \oplus_{\scriptscriptstyle E}, \otimes_{\scriptscriptstyle E})$ turn out to be the well-known geodesics of the Beltrami-Klein ball model of hyperbolic geometry. They are segments of Euclidean straight lines, as shown in Figs. 3.3 – 3.4.

Figure 3.4 presents the gyrosegment AB that joins the points A and B in the Einstein gyrovector plane $(\mathbb{R}^2_{s=1}, \oplus_{\scriptscriptstyle E}, \otimes_{\scriptscriptstyle E})$ along with its gyromidpoint M_{AB}, and a generic point P lying between A and B. Since the points A, B, P are gyrocollinear, they satisfy the gyrotriangle equality, Theorem 3.28, shown in the figure.

Considering the parameter $t \in \mathbb{R}$ of (3.112) as "time", the generic point P of the gyroline L_{AB} in Fig. 3.4 travels along the gyrosegment AB during the time interval $0 \le t \le 1$. It reaches the point A at "time" $t = 0$, it reaches the gyromidpoint M_{AB} at "time" $t = 1/2$, as we see from (3.60), p. 69, and owing to the left cancellation law, it reaches the point B at "time" $t = 1$.

The gyroline equation (3.112) is coordinate free. However, in order to draw graphs of gyrolines, as in Figs. 3.3 – 3.4, we must introduce a coordinate system. Accordingly, we introduce into the unit discs in Figs. 3.3 – 3.4 rectangular Cartesian $x_1 x_2$-coordinates with the condition $x_1^2 + x_2^2 < s^2$ ($s = 1$ in the figures). Relative to the x_1, x_2-coordinates we select, as an example, the points $A = (-0.5, -0.1)$ and $B = (0, -0.5)$ in the unit disc of Fig. 3.3, and the points $A = (-0.5, 0)$ and $B = (0.1, 0.8)$ in the unit disc of Fig. 3.4. We, then, employ Einstein addition $\oplus_{\scriptscriptstyle E}$ and Einstein scalar multiplication $\otimes_{\scriptscriptstyle E}$ for various values of $t \in \mathbb{R}$ in order to calculate points of the gyroline graphs in Figs. 3.3 – 3.4.

An important contact with the elementary differential geometry of the Beltrami-Klein disc model of hyperbolic geometry is provided by the so called *Riemannian gyroline element* of Einstein gyrometric (3.21), p. 61, of the Einstein gyrovector plane $(\mathbb{R}^2_c, \oplus_{\scriptscriptstyle E}, \otimes_{\scriptscriptstyle E})$. Let $\mathbf{v} \in \mathbb{R}^2_c$ be an element of the disc which, relative to the Cartesian x_1, x_2-coordinates of the disc, is written as the column vector $(x_1, x_2)^t$. Accordingly, the differential $d\mathbf{v}$ of $\mathbf{v}$ is written as $d\mathbf{v} = (dx_1, dx_2)^t$. We now consider the *gyrodifferential*

$$d\mathbf{s} = (\mathbf{v} + d\mathbf{v}) \ominus_{\scriptscriptstyle E} \mathbf{v} = \begin{pmatrix} x_1 + dx_1 \\ x_2 + dx_2 \end{pmatrix} \ominus_{\scriptscriptstyle E} \begin{pmatrix} x_1 \\ x_2 \end{pmatrix}, \tag{3.113}$$

the square of which, $d\mathbf{s}^2 = d\mathbf{s} \cdot d\mathbf{s} = \|d\mathbf{s}\|^2$, takes the form

$$d\mathbf{s}^2 = \|(\mathbf{v} + d\mathbf{v}) \ominus_{\scriptscriptstyle E} \mathbf{v}\|^2 = \left\| \begin{pmatrix} x_1 + dx_1 \\ x_2 + dx_2 \end{pmatrix} \ominus_{\scriptscriptstyle E} \begin{pmatrix} x_1 \\ x_2 \end{pmatrix} \right\|^2, \tag{3.114}$$

and has an important gyrogeometric interpretation. Indeed, $d\mathbf{s}^2$ of (3.114) is the squared gyrodistance between the two infinitesimally nearby points $\mathbf{v}$ and $\mathbf{v} + d\mathbf{v}$ in an Einstein gyrovector plane, called a *Riemannian gyroline element*. Remarkably, an explicit calculation of $d\mathbf{s}^2$ in (3.114) reveals that $d\mathbf{s}^2$ is the quadratical differential form

$$d\mathbf{s}^2 = c^2 \frac{dx_1^2 + dx_2^2}{c^2 - r^2} + c^2 \frac{(x_1 dx_1 + x_2 dx_2)^2}{(c^2 - r^2)^2}, \tag{3.115}$$

where $r^2 = x_1^2 + x_2^2$.

What is remarkable in the Riemannian gyroline element (3.115) of Einstein gyrometric $d(\mathbf{u}, \mathbf{v}) = \|\mathbf{u} \ominus_{_E} \mathbf{v}\|$ is that it turns out to be identical with the well-known Riemannian line element of the hyperbolic metric of the Beltrami-Klein disc model of hyperbolic geometry. It is presented, for instance, in McCleary [30, p. 220], for $n = 2$, and in Cannon *et al.* [4, ds_K^2, p. 71], for $n \geq 2$. The Riemannian gyroline element of Einstein gyrometric along with its natural generalization to higher dimensions is studied in detail in [61, Sec. 7.6].

We clearly see in this book that if we use our gyrolanguage and gyrovector space theoretic techniques, many familiar equations of hyperbolic geometry take on unexpected grace. A point in case is the Riemannian gyroline element which turns out to coincide with the Riemannian line element of various models of hyperbolic geometry. Cannon *et al.* admit in [4, p. 71] that the Riemannian line element $d\mathbf{s}^2$ in (3.115) of the Beltrami-Klein model of hyperbolic geometry "has always struck us as particularly ugly and unintuitive". Readers of this book are likely to disagree with Cannon *et al.*, since the Riemannian line element $d\mathbf{s}^2$ in (3.115) takes on unexpected grace and obvious meaning with respect to the Riemannian line element $d\mathbf{s}^2$ in (3.114) from which it is naturally derived. More about the contact that the gyrodistance notion and the Riemannian gyroline element in gyrovector spaces make with classical notions in differential geometry is found in [61, Chap. 7] and [57].

3.10   EINSTEIN GYROMIDPOINTS AND GYROTRIANGLE GYROCENTROIDS

Analogies that gyromidpoints in gyrovector spaces share with midpoints in vector spaces were studied in Sec. 3.3. In the context of Einstein gyrovector spaces gyromidpoints have a relativistic mechanical interpretation, giving rise to further analogies with the common classical mechanical interpretation that midpoints in vector spaces possess.

Thus, for instance, if two equal masses, m, are situated at each of the two points $\mathbf{u}$ and $\mathbf{v}$ of a Newtonian velocity vector space, then their well-known center of momentum is identical with their midpoint, as we see from (3.122), p. 87. In full analogy, if two equal masses, m, are situated at each of the two points $\mathbf{u}$ and $\mathbf{v}$ of an Einsteinian velocity gyrovector space, then their well-known center of momentum is identical with their gyromidpoint, as we see from (3.124), p. 88.

Theorem 3.33.   (The Einstein Gyromidpoint). *Let* $\mathbf{u}, \mathbf{v} \in \mathbb{V}_s$ *be any two points of an Einstein gyrovector space* $(\mathbb{V}_s, \oplus_{_E}, \otimes_{_E})$. *The gyromidpoint* $\mathbf{m}_{\mathbf{uv}}$ *of the points* $\mathbf{u}$ *and* $\mathbf{v}$ *is given by the equation*

$$\mathbf{m}_{\mathbf{uv}} = \frac{\gamma_{\mathbf{u}} \mathbf{u} + \gamma_{\mathbf{v}} \mathbf{v}}{\gamma_{\mathbf{u}} + \gamma_{\mathbf{v}}}. \tag{3.116}$$

Proof. The gyromidpoint $\mathbf{m}_{\mathbf{uv}}$ of two points $\mathbf{u}$ and $\mathbf{v}$ of the abstract gyrovector space $(G, \oplus, \otimes)$ is given by (3.66), p. 70, implying

$$\mathbf{m}_{\mathbf{uv}} = \tfrac{1}{2} \otimes_{_E} (\mathbf{u} \boxplus_{_E} \mathbf{v}) \tag{3.117}$$

in the Einstein gyrovector space $(\mathbb{V}_s, \oplus_{_E}, \otimes_{_E})$.

It follows from (3.117), by (2.89), p. 49, and by the scalar associative law of gyrovector spaces, that the gyromidpoint $\mathbf{m_{uv}}$ in Einstein gyrovector spaces is given by

$$
\begin{aligned}
\mathbf{m_{uv}} &= \tfrac{1}{2} \otimes_{\scriptscriptstyle E} (\mathbf{u} \boxplus_{\scriptscriptstyle E} \mathbf{v}) \\
&= \tfrac{1}{2} \otimes_{\scriptscriptstyle E} \{2 \otimes_{\scriptscriptstyle E} \frac{\gamma_{\mathbf{u}} \mathbf{u} + \gamma_{\mathbf{v}} \mathbf{v}}{\gamma_{\mathbf{u}} + \gamma_{\mathbf{v}}}\} \\
&= \frac{\gamma_{\mathbf{u}} \mathbf{u} + \gamma_{\mathbf{v}} \mathbf{v}}{\gamma_{\mathbf{u}} + \gamma_{\mathbf{v}}},
\end{aligned}
\tag{3.118}
$$

as desired. □

Gyromidpoints give rise to gyrotriangle gyromedians in gyrovector spaces, in analogy with triangle medians in vector spaces, as we see in the following definition.

Definition 3.34.   (Gyrotriangle Gyromedians, Gyrocentroids). *The gyrosegment connecting the gyromidpoint of a side of a gyrotriangle with its opposite vertex, Figs. 3.5 and 3.7, is called a gyromedian. The point of concurrency of the three gyrotriangle gyromedians is called the gyrotriangle gyrocentroid.*

The gyromidpoints of the three sides of a gyrotriangle $\mathbf{uvw}$ in an Einstein (Möbius) gyrovector plane and its gyrocentroid are shown in Fig. 3.5 (Fig. 3.7, respectively).

Gyrolines in an Einstein gyrovector spaces are segments of Euclidean straight lines. Hence, by elementary methods of linear algebra, one can verify that the three gyromedians in Fig. 3.5 are concurrent, and that the point of concurrency, that is, the gyrocentroid $\mathbf{m_{uvw}}$ of gyrotriangle $\mathbf{uvw}$, is given by the elegant equation (3.119) that we place in the following theorem.

Theorem 3.35.   (The Einstein Gyrotriangle Gyrocentroid). *Let $\mathbf{u}, \mathbf{v}, \mathbf{w} \in \mathbb{R}_s^n$ be any three nongyrocollinear points of an Einstein gyrovector space $(\mathbb{R}_s^n, \oplus_{\scriptscriptstyle E}, \otimes_{\scriptscriptstyle E})$. The gyrocentroid $\mathbf{m_{uvw}}$ of the gyrotriangle $\mathbf{uvw}$, Fig. 3.5, is given by the equation*

$$
\mathbf{m_{uvw}} = \frac{\gamma_{\mathbf{u}} \mathbf{u} + \gamma_{\mathbf{v}} \mathbf{v} + \gamma_{\mathbf{w}} \mathbf{w}}{\gamma_{\mathbf{u}} + \gamma_{\mathbf{v}} + \gamma_{\mathbf{w}}}.
\tag{3.119}
$$

Proof. Gyrosegments in an Einstein gyrovector space $\mathbb{R}_s^n$ are Euclidean segments of straight lines in $\mathbb{R}^n$. Hence, by methods of linear algebra one can show that the gyrosegments $\mathbf{um_{vw}}$, $\mathbf{vm_{uw}}$, and $\mathbf{wm_{uv}}$ in Fig. 3.5 are concurrent, the concurrency point being $\mathbf{m_{uvw}}$ in (3.119). □

Gyromidpoints are gyrocovariant, as shown in (3.72), p. 72. Therefore a left gyrotranslation by $\mathbf{x} \in \mathbb{R}_s^2$ of the gyrotriangle $\mathbf{uvw}$ in Fig. 3.5 into the gyrotriangle $(\mathbf{x} \oplus \mathbf{u})(\mathbf{x} \oplus \mathbf{v})(\mathbf{x} \oplus \mathbf{w})$ in Fig. 3.6

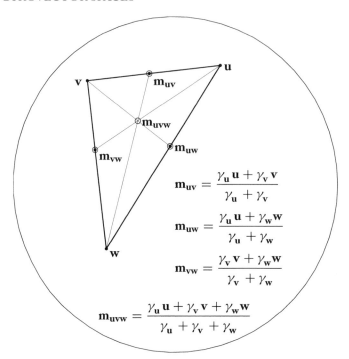

$$\mathbf{m}_{\mathbf{uv}} = \frac{\gamma_{\mathbf{u}}\mathbf{u} + \gamma_{\mathbf{v}}\mathbf{v}}{\gamma_{\mathbf{u}} + \gamma_{\mathbf{v}}}$$

$$\mathbf{m}_{\mathbf{uw}} = \frac{\gamma_{\mathbf{u}}\mathbf{u} + \gamma_{\mathbf{w}}\mathbf{w}}{\gamma_{\mathbf{u}} + \gamma_{\mathbf{w}}}$$

$$\mathbf{m}_{\mathbf{vw}} = \frac{\gamma_{\mathbf{v}}\mathbf{v} + \gamma_{\mathbf{w}}\mathbf{w}}{\gamma_{\mathbf{v}} + \gamma_{\mathbf{w}}}$$

$$\mathbf{m}_{\mathbf{uvw}} = \frac{\gamma_{\mathbf{u}}\mathbf{u} + \gamma_{\mathbf{v}}\mathbf{v} + \gamma_{\mathbf{w}}\mathbf{w}}{\gamma_{\mathbf{u}} + \gamma_{\mathbf{v}} + \gamma_{\mathbf{w}}}$$

Figure 3.5: The gyromidpoints $\mathbf{m}_{\mathbf{uv}}$, $\mathbf{m}_{\mathbf{vw}}$, and $\mathbf{m}_{\mathbf{uw}}$ of the three sides, $\mathbf{uv}$, $\mathbf{vw}$, and $\mathbf{wu}$, of a gyrotriangle $\mathbf{uvw}$ in the Einstein gyrovector space $(\mathbb{V}_s, \oplus_{\mathrm{E}}, \otimes_{\mathrm{E}})$ are shown here for $\mathbb{V}_s = \mathbb{R}_s^2$, along with its gyromedians and its gyrocentroid $\mathbf{m}_{\mathbf{uvw}}$. Interestingly, Einsteinian gyromidpoints and gyrocentroids have interpretation in relativistic mechanics that is fully analogous to the interpretation of Euclidean midpoints and centroids in classical mechanics.

does not distort the gyrotriangle side gyromidpoints and the gyrotriangle gyrocentroid. Hence, the gyrocentroid of the gyrotranslated gyrotriangle $(\mathbf{x}\oplus\mathbf{u})(\mathbf{x}\oplus\mathbf{v})(\mathbf{x}\oplus\mathbf{w})$, Fig. 3.6, is given by the equation

$$\mathbf{x}\oplus\mathbf{m}_{\mathbf{uvw}} = \mathbf{m}_{(\mathbf{x}\oplus\mathbf{u})(\mathbf{x}\oplus\mathbf{v})(\mathbf{x}\oplus\mathbf{w})} = \frac{\gamma_{\mathbf{x}\oplus\mathbf{u}}(\mathbf{x}\oplus\mathbf{u}) + \gamma_{\mathbf{x}\oplus\mathbf{v}}(\mathbf{x}\oplus\mathbf{v}) + \gamma_{\mathbf{x}\oplus\mathbf{w}}(\mathbf{x}\oplus\mathbf{w})}{\gamma_{\mathbf{x}\oplus\mathbf{u}} + \gamma_{\mathbf{x}\oplus\mathbf{v}} + \gamma_{\mathbf{x}\oplus\mathbf{w}}}, \tag{3.120}$$

thus uncovering the interesting identity

$$\mathbf{x}\oplus\frac{\gamma_{\mathbf{u}}\mathbf{u} + \gamma_{\mathbf{v}}\mathbf{v} + \gamma_{\mathbf{w}}\mathbf{w}}{\gamma_{\mathbf{u}} + \gamma_{\mathbf{v}} + \gamma_{\mathbf{w}}} = \frac{\gamma_{\mathbf{x}\oplus\mathbf{u}}(\mathbf{x}\oplus\mathbf{u}) + \gamma_{\mathbf{x}\oplus\mathbf{v}}(\mathbf{x}\oplus\mathbf{v}) + \gamma_{\mathbf{x}\oplus\mathbf{w}}(\mathbf{x}\oplus\mathbf{w})}{\gamma_{\mathbf{x}\oplus\mathbf{u}} + \gamma_{\mathbf{x}\oplus\mathbf{v}} + \gamma_{\mathbf{x}\oplus\mathbf{w}}} \tag{3.121}$$

in any Einstein gyrovector space. Identity (3.121) presents a remarkable harmonious interplay between Einstein addition $\oplus$ and the common vector addition $+$ and the common addition of real numbers, also denoted by $+$.

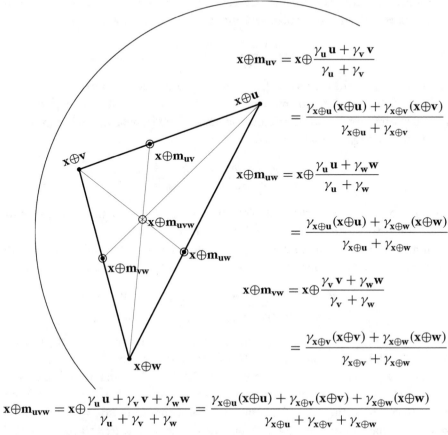

$$\mathbf{x} \oplus \mathbf{m_{uv}} = \mathbf{x} \oplus \frac{\gamma_\mathbf{u}\mathbf{u} + \gamma_\mathbf{v}\mathbf{v}}{\gamma_\mathbf{u} + \gamma_\mathbf{v}}$$

$$= \frac{\gamma_{\mathbf{x} \oplus \mathbf{u}}(\mathbf{x} \oplus \mathbf{u}) + \gamma_{\mathbf{x} \oplus \mathbf{v}}(\mathbf{x} \oplus \mathbf{v})}{\gamma_{\mathbf{x} \oplus \mathbf{u}} + \gamma_{\mathbf{x} \oplus \mathbf{v}}}$$

$$\mathbf{x} \oplus \mathbf{m_{uw}} = \mathbf{x} \oplus \frac{\gamma_\mathbf{u}\mathbf{u} + \gamma_\mathbf{w}\mathbf{w}}{\gamma_\mathbf{u} + \gamma_\mathbf{w}}$$

$$= \frac{\gamma_{\mathbf{x} \oplus \mathbf{u}}(\mathbf{x} \oplus \mathbf{u}) + \gamma_{\mathbf{x} \oplus \mathbf{w}}(\mathbf{x} \oplus \mathbf{w})}{\gamma_{\mathbf{x} \oplus \mathbf{u}} + \gamma_{\mathbf{x} \oplus \mathbf{w}}}$$

$$\mathbf{x} \oplus \mathbf{m_{vw}} = \mathbf{x} \oplus \frac{\gamma_\mathbf{v}\mathbf{v} + \gamma_\mathbf{w}\mathbf{w}}{\gamma_\mathbf{v} + \gamma_\mathbf{w}}$$

$$= \frac{\gamma_{\mathbf{x} \oplus \mathbf{v}}(\mathbf{x} \oplus \mathbf{v}) + \gamma_{\mathbf{x} \oplus \mathbf{w}}(\mathbf{x} \oplus \mathbf{w})}{\gamma_{\mathbf{x} \oplus \mathbf{v}} + \gamma_{\mathbf{x} \oplus \mathbf{w}}}$$

$$\mathbf{x} \oplus \mathbf{m_{uvw}} = \mathbf{x} \oplus \frac{\gamma_\mathbf{u}\mathbf{u} + \gamma_\mathbf{v}\mathbf{v} + \gamma_\mathbf{w}\mathbf{w}}{\gamma_\mathbf{u} + \gamma_\mathbf{v} + \gamma_\mathbf{w}} = \frac{\gamma_{\mathbf{x} \oplus \mathbf{u}}(\mathbf{x} \oplus \mathbf{u}) + \gamma_{\mathbf{x} \oplus \mathbf{v}}(\mathbf{x} \oplus \mathbf{v}) + \gamma_{\mathbf{x} \oplus \mathbf{w}}(\mathbf{x} \oplus \mathbf{w})}{\gamma_{\mathbf{x} \oplus \mathbf{u}} + \gamma_{\mathbf{x} \oplus \mathbf{v}} + \gamma_{\mathbf{x} \oplus \mathbf{w}}}$$

Figure 3.6: A left gyrotranslation of the gyrotriangle **uvw** of Fig. 3.5 by **x** into gyrotriangle $(\mathbf{x} \oplus \mathbf{u})(\mathbf{x} \oplus \mathbf{v})(\mathbf{x} \oplus \mathbf{w})$ is shown here. It results in corresponding left gyrotranslations by **x** of the gyrotriangle gyromidpoints and the gyrotriangle gyrocentroid as well. We see here a remarkably harmonious interplay between vector addition + and scalar addition + and scalar multiplication between scalars and vectors in the inner product space $\mathbb{V}$ on the one hand, and Einstein addition $\oplus_E = \oplus$ in the Einstein gyrovector space of the ball $\mathbb{V}_s = (\mathbb{V}_s, \oplus, \otimes)$ on the other hand.

Euclidean midpoints and Euclidean triangle centroids have an interesting classical mechanical interpretation [22]. In classical mechanics, the center of momentum of a system of two particles with equal masses, $m > 0$, and Newtonian velocities **u** and **v**, where $\mathbf{u}, \mathbf{v} \in \mathbb{R}^3$, is

$$\frac{m\mathbf{u} + m\mathbf{v}}{2m} = \frac{\mathbf{u} + \mathbf{v}}{2}, \tag{3.122}$$

which turns out to be the Euclidean midpoint, $\mathbf{m}_{\mathbf{uv}}^e$, of **u** and **v** in $\mathbb{R}^3$.

Similarly, in classical mechanics, the center of momentum of a system of three particles with equal masses, $m > 0$, and Newtonian velocities $\mathbf{u}$, $\mathbf{v}$ and $\mathbf{w}$, $\mathbf{u}, \mathbf{v}, \mathbf{w} \in \mathbb{R}^3$, is

$$\frac{m\mathbf{u} + m\mathbf{v} + m\mathbf{w}}{3m} = \frac{\mathbf{u} + \mathbf{v} + \mathbf{w}}{3} , \qquad (3.123)$$

which turns out to be the centroid, $\mathbf{m}_{\mathbf{uvw}}^e$, of the Euclidean triangle $\mathbf{uvw}$.

Interestingly, in relativistic mechanics, the center of momentum of a system of two particles with equal masses, $m > 0$, and Einsteinian velocities (that is, relativistically admissible velocities) $\mathbf{u}$ and $\mathbf{v}$, where $\mathbf{u}, \mathbf{v} \in \mathbb{R}_c^3$, is, Figs, 3.5 – 3.6,

$$\frac{m\gamma_{\mathbf{u}}\mathbf{u} + m\gamma_{\mathbf{v}}\mathbf{v}}{m\gamma_{\mathbf{u}} + m\gamma_{\mathbf{v}}} = \frac{\gamma_{\mathbf{u}}\mathbf{u} + \gamma_{\mathbf{v}}\mathbf{v}}{\mathbf{u} + \mathbf{v}} , \qquad (3.124)$$

which turns out to be the gyromidpoint, $\mathbf{m}_{\mathbf{uv}}$, of $\mathbf{u}$ and $\mathbf{v}$ in the Einstein gyrovector space $(\mathbb{R}_c^3, \oplus_{\mathrm{E}}, \otimes_{\mathrm{E}})$, as we see from Identity (3.116) of Theorem 3.33.

Similarly, in relativistic mechanics, the center of momentum of a system of three particles with equal masses, m, and Einsteinian velocities $\mathbf{u}$, $\mathbf{v}$, and $\mathbf{w}$, where $\mathbf{u}, \mathbf{v}, \mathbf{w} \in \mathbb{R}_c^3$, is

$$\frac{m\gamma_{\mathbf{u}}\mathbf{u} + m\gamma_{\mathbf{v}}\mathbf{v} + m\gamma_{\mathbf{w}}\mathbf{w}}{m\gamma_{\mathbf{u}} + m\gamma_{\mathbf{v}} + m\gamma_{\mathbf{w}}} = \frac{\gamma_{\mathbf{u}}\mathbf{u} + \gamma_{\mathbf{v}}\mathbf{v} + \gamma_{\mathbf{w}}\mathbf{w}}{\gamma_{\mathbf{u}} + \gamma_{\mathbf{v}} + \gamma_{\mathbf{w}}} , \qquad (3.125)$$

which turns out to be the gyrocentroid, $\mathbf{m}_{\mathbf{uvw}}$, of the gyrotriangle $\mathbf{uvw}$ in the Einstein gyrovector space $(\mathbb{R}_c^3, \oplus_{\mathrm{E}}, \otimes_{\mathrm{E}})$, as explained in (3.119) and shown in Fig. 3.5.

The analogies that the classical results in (3.122) – (3.123) share with their modern counterparts in (3.124) – (3.125) thus extend their validity from analogies between Euclidean and hyperbolic geometry to corresponding analogies between classical and relativistic mechanics. We thus see here not only mathematical analogies in action, but also the play of *"Analogies Between Analogies"*, the latter being the title of an interesting book by the famous mathematician Stanislaw Ulam (1909 – 1984) [51].

3.11   MÖBIUS GYROTRIANGLE GYROMEDIANS AND GYROCENTROIDS

Let $G_e = (\mathbb{V}_s, \oplus_{\mathrm{E}}, \otimes_{\mathrm{E}})$ and $G_m = (\mathbb{V}_s, \oplus_{\mathrm{M}}, \otimes_{\mathrm{M}})$ be, respectively, the Einstein and the Möbius gyrovector spaces of the ball $\mathbb{V}_s$ of a real inner product space $\mathbb{V}$. They are isomorphic, with the isomorphism ϕ_{EM} from G_m into G_e, and its inverse isomorphism ϕ_{ME} from G_e into G_m, given by (2.99) and (2.101), p. 51. Accordingly, the correspondence between elements of G_m and G_e is given by the equations

$$\mathbf{v}_e = \phi_{\mathrm{EM}}\mathbf{v}_m = 2\otimes_{\mathrm{M}}\mathbf{v}_m$$

$$\mathbf{v}_m = \phi_{\mathrm{ME}}\mathbf{v}_e = \tfrac{1}{2}\otimes_{\mathrm{E}}\mathbf{v}_e = \frac{\gamma_{\mathbf{v}_e}}{1 + \gamma_{\mathbf{v}_e}}\mathbf{v}_e , \qquad (3.126)$$

$\mathbf{v}_e \in G_e, \mathbf{v}_m \in G_m$, noting the Einstein half (3.106), p. 81.

Hence, by (3.126) and the first identity in (3.111) we have

$$\gamma_{\mathbf{v}_e} = \gamma_{2\otimes_M \mathbf{v}_m}$$
$$= 2\gamma_{\mathbf{v}_m}^2 - 1 . \tag{3.127}$$

Similarly, following (3.126) and (3.109) we have,

$$\gamma_{\mathbf{v}_e} \mathbf{v}_e = \gamma_{2\otimes_M \mathbf{v}_m} 2\otimes_M \mathbf{v}_m$$
$$= 2\gamma_{\mathbf{v}_m}^2 \mathbf{v}_m . \tag{3.128}$$

Hence, by (3.116), (3.127), and (3.128) we have

$$\mathbf{m}_{\mathbf{u}_e \mathbf{v}_e} = \frac{\gamma_{\mathbf{u}_e} \mathbf{u}_e + \gamma_{\mathbf{v}_e} \mathbf{v}_e}{\gamma_{\mathbf{u}_e} + \gamma_{\mathbf{v}_e}}$$

$$= \frac{2\gamma_{\mathbf{u}_m}^2 \mathbf{u}_m + 2\gamma_{\mathbf{v}_m}^2 \mathbf{v}_m}{(2\gamma_{\mathbf{u}_m}^2 - 1) + (2\gamma_{\mathbf{v}_m}^2 - 1)} \tag{3.129}$$

$$= \frac{\gamma_{\mathbf{u}_m}^2 \mathbf{u}_m + \gamma_{\mathbf{v}_m}^2 \mathbf{v}_m}{\gamma_{\mathbf{u}_m}^2 + \gamma_{\mathbf{v}_m}^2 - 1} ,$$

so that

$$\mathbf{m}_{\mathbf{u}_m \mathbf{v}_m} = \tfrac{1}{2} \otimes_E \mathbf{m}_{\mathbf{u}_e \mathbf{v}_e}$$
$$= \tfrac{1}{2} \otimes_E \frac{\gamma_{\mathbf{u}_m}^2 \mathbf{u}_m + \gamma_{\mathbf{v}_m}^2 \mathbf{v}_m}{\gamma_{\mathbf{u}_m}^2 + \gamma_{\mathbf{v}_m}^2 - 1} . \tag{3.130}$$

Möbius gyromidpoints, gyromedians, and a gyrocentroid, are shown in Fig. 3.7. Formalizing the result in (3.130) for the Möbius gyromidpoint, we have the following theorem.

Theorem 3.36.   (The Möbius Gyromidpoint). *Let $\mathbf{u}, \mathbf{v} \in \mathbb{V}_s$ be any two points of a Möbius gyrovector space $(\mathbb{V}_s, \oplus_M, \otimes_M)$. The gyromidpoint $\mathbf{m}_{\mathbf{uv}}$ of the gyrosegment $\mathbf{uv}$ joining the points $\mathbf{u}$ and $\mathbf{v}$ in $\mathbb{V}_s$ is given by the equation*

$$\mathbf{m}_{\mathbf{uv}} = \tfrac{1}{2} \otimes_M \frac{\gamma_{\mathbf{u}}^2 \mathbf{u} + \gamma_{\mathbf{v}}^2 \mathbf{v}}{\gamma_{\mathbf{u}}^2 + \gamma_{\mathbf{v}}^2 - 1} . \tag{3.131}$$

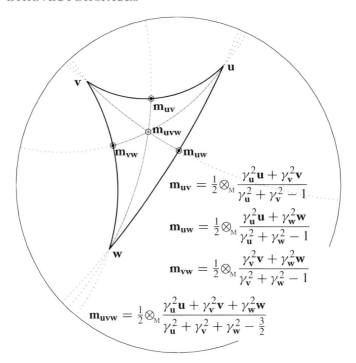

$$\mathbf{m_{uv}} = \tfrac{1}{2}\otimes_{\mathrm{M}}\frac{\gamma_{\mathbf{u}}^2\mathbf{u}+\gamma_{\mathbf{v}}^2\mathbf{v}}{\gamma_{\mathbf{u}}^2+\gamma_{\mathbf{v}}^2-1}$$

$$\mathbf{m_{uw}} = \tfrac{1}{2}\otimes_{\mathrm{M}}\frac{\gamma_{\mathbf{u}}^2\mathbf{u}+\gamma_{\mathbf{w}}^2\mathbf{w}}{\gamma_{\mathbf{u}}^2+\gamma_{\mathbf{w}}^2-1}$$

$$\mathbf{m_{vw}} = \tfrac{1}{2}\otimes_{\mathrm{M}}\frac{\gamma_{\mathbf{v}}^2\mathbf{v}+\gamma_{\mathbf{w}}^2\mathbf{w}}{\gamma_{\mathbf{v}}^2+\gamma_{\mathbf{w}}^2-1}$$

$$\mathbf{m_{uvw}} = \tfrac{1}{2}\otimes_{\mathrm{M}}\frac{\gamma_{\mathbf{u}}^2\mathbf{u}+\gamma_{\mathbf{v}}^2\mathbf{v}+\gamma_{\mathbf{w}}^2\mathbf{w}}{\gamma_{\mathbf{u}}^2+\gamma_{\mathbf{v}}^2+\gamma_{\mathbf{w}}^2-\tfrac{3}{2}}$$

Figure 3.7: The gyromidpoints $\mathbf{m_{uv}}$, $\mathbf{m_{vw}}$, and $\mathbf{m_{uw}}$ of the three sides, $\mathbf{uv}$, $\mathbf{vw}$ and $\mathbf{wu}$, of a gyrotriangle $\mathbf{uvw}$ in the Möbius gyrovector space $(\mathbb{V}_s, \oplus_{\mathrm{M}}, \otimes_{\mathrm{M}})$ are shown here for $\mathbb{V}_s = \mathbb{R}_s^2$, along with its gyromedians and its gyrocentroid $\mathbf{m_{uvw}}$.

Similar to (3.129) and (3.130) we have, by (3.119), (3.127), and (3.128),

$$
\begin{aligned}
\mathbf{m_{u_ev_ew_e}} &= \frac{\gamma_{\mathbf{u}_e}\mathbf{u}_e + \gamma_{\mathbf{v}_e}\mathbf{v}_e + \gamma_{\mathbf{w}_e}\mathbf{w}_e}{\gamma_{\mathbf{u}_e} + \gamma_{\mathbf{v}_e} + \gamma_{\mathbf{w}_e}} \\[2mm]
&= \frac{2\gamma_{\mathbf{u}_m}^2\mathbf{u}_m + 2\gamma_{\mathbf{v}_m}^2\mathbf{v}_m + 2\gamma_{\mathbf{w}_m}^2\mathbf{w}_m}{(2\gamma_{\mathbf{u}_m}^2-1)+(2\gamma_{\mathbf{v}_m}^2-1)+(2\gamma_{\mathbf{w}_m}^2-1)} \\[2mm]
&= \frac{\gamma_{\mathbf{u}_m}^2\mathbf{u}_m + \gamma_{\mathbf{v}_m}^2\mathbf{v}_m + \gamma_{\mathbf{w}_m}^2\mathbf{w}_m}{\gamma_{\mathbf{u}_m}^2 + \gamma_{\mathbf{v}_m}^2 + \gamma_{\mathbf{w}_m}^2 - \tfrac{3}{2}},
\end{aligned}
\tag{3.132}
$$

so that

$$
\begin{aligned}
\mathbf{m_{u_mv_mw_m}} &= \tfrac{1}{2}\otimes_{\mathrm{M}}\mathbf{m_{u_ev_ew_e}} \\[2mm]
&= \tfrac{1}{2}\otimes_{\mathrm{M}}\frac{\gamma_{\mathbf{u}_m}^2\mathbf{u}_m + \gamma_{\mathbf{v}_m}^2\mathbf{v}_m + \gamma_{\mathbf{w}_m}^2\mathbf{w}_m}{\gamma_{\mathbf{u}_m}^2 + \gamma_{\mathbf{v}_m}^2 + \gamma_{\mathbf{w}_m}^2 - \tfrac{3}{2}}
\end{aligned}
\tag{3.133}
$$

for any $\mathbf{u}_m, \mathbf{v}_m, \mathbf{w}_m \in G_m$.

Formalizing the result in (3.133), we have the following theorem.

Theorem 3.37.   (The Möbius Gyrotriangle Gyrocentroid). *Let* $\mathbf{u}, \mathbf{v}, \mathbf{w} \in \mathbb{V}_s$ *be any three nongy-rocollinear points of a Möbius gyrovector space* $\mathbb{V}_s = (\mathbb{V}_s, \oplus_M, \otimes_M)$. *The centroid* $\mathbf{m_{uvw}}$, *Fig. 3.7, of gyrotriangle* $\mathbf{uvw}$ *in* $\mathbb{V}_s$ *is given by the equation*

$$\mathbf{m_{uvw}} = \tfrac{1}{2}\otimes_M \frac{\gamma_{\mathbf{u}}^2 \mathbf{u} + \gamma_{\mathbf{v}}^2 \mathbf{v} + \gamma_{\mathbf{w}}^2 \mathbf{w}}{\gamma_{\mathbf{u}}^2 + \gamma_{\mathbf{v}}^2 + \gamma_{\mathbf{w}}^2 - \tfrac{3}{2}} . \tag{3.134}$$

The results (3.131) and (3.134) involve formulas that are more complex than their counterparts (3.116) and (3.119) in Einstein gyrovector spaces. One may readily calculate explicitly the scalar multiplication by $(1/2)\otimes_M$ in (3.131) and (3.134). However, in this case the resulting formulas will appear more complex. It is thus clear that rather than calculating the results (3.131) and (3.134) directly, it is simpler to obtain them by translating their Einsteinian counterparts from Einstein gyrovector spaces to Möbius gyrovector spaces, as we did in this section.

3.12   THE GYROPARALLELOGRAM

Definition 3.38.   (Gyroparallelograms). *Let* $\mathbf{a}, \mathbf{b}$, *and* $\mathbf{c}$ *be any three points in a gyrovector space* $(G, \oplus, \otimes)$. *Then, the four points* $\mathbf{a}, \mathbf{b}, \mathbf{c}, \mathbf{d}$ *in* G *are the vertices of the gyroparallelogram* $\mathbf{abdc}$, *ordered either clockwise of counterclockwise, Fig. 3.8, if* $\mathbf{d}$ *satisfies the gyroparallelogram condition (2.13), p. 35,*

$$\mathbf{d} = (\mathbf{b} \boxplus \mathbf{c})\ominus\mathbf{a} . \tag{3.135}$$

The gyroparallelogram is degenerate if the three points $\mathbf{a}, \mathbf{b}$, *and* $\mathbf{c}$ *are gyrocollinear.*

If the gyroparallelogram $\mathbf{abdc}$ *is non-degenerate, then the two vertices in each of the pairs* $(\mathbf{a}, \mathbf{d})$ *and* $(\mathbf{b}, \mathbf{c})$ *are said to be opposite to one another. The gyrosegments of adjacent vertices,* $\mathbf{ab}, \mathbf{bd}, \mathbf{dc}$, *and* $\mathbf{ca}$, *are the sides of the gyroparallelogram. The gyrosegments* $\mathbf{ad}$ *and* $\mathbf{bc}$ *that link opposite vertices of the non-degenerate gyroparallelogram* $\mathbf{abdc}$ *are the gyrodiagonals of the gyroparallelogram.*

The gyrocenter $\mathbf{m_{abdc}}$ *of the gyroparallelogram* $\mathbf{abdc}$ *is the gyromidpoint of each of its two gyrodiagonals, so that* $\mathbf{m_{abdc}} = \mathbf{m_{ad}} = \mathbf{m_{bc}}$ *(see Theorem 3.39 below).*

In what seemingly sounds like a contradiction in terms we have extended in Def. 3.38 the Euclidean parallelogram into hyperbolic geometry where the parallel postulate is denied. The resulting gyroparallelogram shares remarkable analogies with its Euclidean counterpart, giving rise to the gyroparallelogram law of gyrovector addition, which is fully analogous to the common parallelogram law of vector addition in Euclidean geometry. A gyroparallelogram in an Einstein gyrovector plane, that is, in the Beltrami-Klein disc model of hyperbolic geometry, is presented in Fig. 3.8. A gyroparallelogram in a Möbius gyrovector plane, that is, in the Poincaré disc model of hyperbolic geometry, is presented in Fig. 3.11, p. 99.

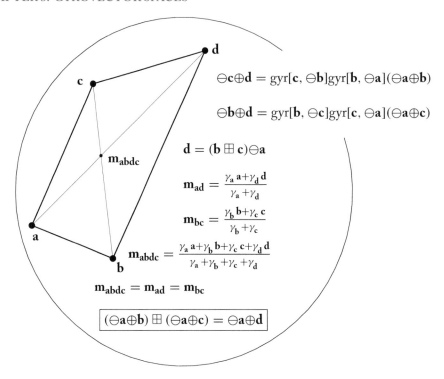

$$\ominus \mathbf{c} \oplus \mathbf{d} = \text{gyr}[\mathbf{c}, \ominus \mathbf{b}]\text{gyr}[\mathbf{b}, \ominus \mathbf{a}](\ominus \mathbf{a} \oplus \mathbf{b})$$

$$\ominus \mathbf{b} \oplus \mathbf{d} = \text{gyr}[\mathbf{b}, \ominus \mathbf{c}]\text{gyr}[\mathbf{c}, \ominus \mathbf{a}](\ominus \mathbf{a} \oplus \mathbf{c})$$

$$\mathbf{d} = (\mathbf{b} \boxplus \mathbf{c}) \ominus \mathbf{a}$$

$$\mathbf{m}_{ad} = \frac{\gamma_a \mathbf{a} + \gamma_d \mathbf{d}}{\gamma_a + \gamma_d}$$

$$\mathbf{m}_{bc} = \frac{\gamma_b \mathbf{b} + \gamma_c \mathbf{c}}{\gamma_b + \gamma_c}$$

$$\mathbf{m}_{abdc} = \frac{\gamma_a \mathbf{a} + \gamma_b \mathbf{b} + \gamma_c \mathbf{c} + \gamma_d \mathbf{d}}{\gamma_a + \gamma_b + \gamma_c + \gamma_d}$$

$$\mathbf{m}_{abdc} = \mathbf{m}_{ad} = \mathbf{m}_{bc}$$

$$\boxed{(\ominus \mathbf{a} \oplus \mathbf{b}) \boxplus (\ominus \mathbf{a} \oplus \mathbf{c}) = \ominus \mathbf{a} \oplus \mathbf{d}}$$

Figure 3.8: The Einstein gyroparallelogram. An Einstein gyroparallelogram is a gyroparallelogram, Def. 3.38, in an Einstein gyrovector space. Let $\mathbf{a}$, $\mathbf{b}$, $\mathbf{c}$ be any three nongyrocollinear points in an Einstein gyrovector space $(\mathbb{V}_s, \oplus, \otimes)$, and let $\mathbf{d}$ be a point in $\mathbb{V}_s$ given by the gyroparallelogram condition, $\mathbf{d} = (\mathbf{b} \boxplus \mathbf{c}) \ominus \mathbf{a}$. Then the four points $\mathbf{a}$, $\mathbf{b}$, $\mathbf{c}$, $\mathbf{d}$ are the vertices of the Einstein gyroparallelogram **abdc** and, by Theorem 3.42, opposite sides are equal modulo gyrations. A gyroquadrilateral **abdc** in a gyrovector space is a gyroparallelogram if and only if its two gyrodiagonals **ad** and **bc** intersect at their gyromidpoints $\mathbf{m}_{ad}$ and $\mathbf{m}_{bc}$, giving rise to the gyroparallelogram gyrocenter $\mathbf{m}_{abdc}$. Indeed, accordingly, $\mathbf{m}_{abdc} = \mathbf{m}_{ad} = \mathbf{m}_{bc}$. The boxed identity gives rise to the gyroparallelogram law of gyrovector addition, shown in Fig. 3.10.

Theorem 3.39.   (Gyroparallelogram Symmetries). *Every vertex of the gyroparallelogram* **abdc**, *Fig. 3.8, satisfies the gyroparallelogram condition,* (3.135), *that is,*

$$\begin{aligned}
\mathbf{a} &= (\mathbf{b} \boxplus \mathbf{c}) \ominus \mathbf{d} \\
\mathbf{b} &= (\mathbf{a} \boxplus \mathbf{d}) \ominus \mathbf{c} \\
\mathbf{c} &= (\mathbf{a} \boxplus \mathbf{d}) \ominus \mathbf{b} \\
\mathbf{d} &= (\mathbf{b} \boxplus \mathbf{c}) \ominus \mathbf{a} \, .
\end{aligned} \tag{3.136}$$

Furthermore, the two gyrodiagonals of a non-degenerate gyroparallelogram are concurrent, the concurrency point being the gyromidpoint of each of the two gyrodiagonals.

Proof. The last equation in (3.136) is valid by Def. 3.38 of the gyroparallelogram. By the right cancellation law (1.72), p. 19, this equation is equivalent to the equation

$$\mathbf{a} \boxplus \mathbf{d} = \mathbf{b} \boxplus \mathbf{c} .\tag{3.137}$$

By Theorem 2.3, p. 34, the coaddition $\boxplus$ in gyrovector spaces is commutative. Hence, by the right cancellation law (1.71), p. 19, the equation in (3.137) is equivalent to each of the equations in (3.136), thus verifying the first part of the theorem.

Equation (3.137) implies

$$\tfrac{1}{2} \otimes (\mathbf{a} \boxplus \mathbf{d}) = \tfrac{1}{2} \otimes (\mathbf{b} \boxplus \mathbf{c}) .\tag{3.138}$$

By Theorem 3.26, p. 70, the left- and the right-hand side of (3.138) are, respectively, the gyromidpoint of the gyrodiagonal **ad** and the gyrodiagonal **bc**. Hence, (3.138) implies that the gyromidpoints of the two gyrodiagonals of the gyroparallelogram coincide, thus verifying the second part of the theorem. □

A gyroquadrilateral is a not self intersecting gyropolygon with four sides and four vertices in a gyrovector space. It is non-degenerate if any three of its vertices are non-gyrocollinear. The following theorem characterizes non-degenerate gyroquadrilaterals which are non-degenerate gyroparallelograms.

Theorem 3.40. *A non-degenerate gyroquadrilateral* **abdc** *in a gyrovector space is a non-degenerate gyroparallelogram if and only if its gyrodiagonals* **ad** *and* **bc** *intersect at their gyromidpoints.*

Proof. The gyrodiagonals **ad** and **bc** of a non-degenerate gyroquadrilateral **abdc** in a gyrovector space intersect at their gyromidpoints if and only if (3.138) is satisfied. The latter, in turn, is satisfied if and only if the gyroparallelogram condition (3.135) is satisfied, as explained in the proof of Theorem 3.39. Finally, by Def. 3.38, the gyroparallelogram condition is satisfied if and only if the gyroquadrilateral **abdc** is a gyroparallelogram. □

Theorem 3.41. (The Gyroparallelogram (Addition) Law). *Let* **abdc** *be a gyroparallelogram in a gyrovector space* $(G, \oplus, \otimes)$, *Fig. 3.8. Then*

$$(\ominus \mathbf{a} \oplus \mathbf{b}) \boxplus (\ominus \mathbf{a} \oplus \mathbf{c}) = \ominus \mathbf{a} \oplus \mathbf{d} .\tag{3.139}$$

Proof. By Identity (2.40) of Theorem 2.15, p. 39, and by the gyroparallelogram condition (3.135) we have

$$(\ominus \mathbf{a} \oplus \mathbf{b}) \boxplus (\ominus \mathbf{a} \oplus \mathbf{c}) = \ominus \mathbf{a} \oplus \{(\mathbf{b} \boxplus \mathbf{c}) \ominus \mathbf{a}\} = \ominus \mathbf{a} \oplus \mathbf{d} .\tag{3.140}$$

□

In the next theorem we will uncover the relationship between opposite sides of the gyroparallelogram.

Theorem 3.42. *Opposite sides of a gyroparallelogram* **abdc** *in a gyrovector space* $(G, \oplus, \otimes)$ *are equal modulo gyrations, Fig. 3.8,*

$$\ominus\mathbf{c}\oplus\mathbf{d} = \text{gyr}[\mathbf{c}, \ominus\mathbf{b}]\text{gyr}[\mathbf{b}, \ominus\mathbf{a}](\ominus\mathbf{a}\oplus\mathbf{b}) = \text{gyr}[\mathbf{c}, \ominus\mathbf{b}](\mathbf{b}\ominus\mathbf{a})$$
$$\ominus\mathbf{b}\oplus\mathbf{d} = \text{gyr}[\mathbf{b}, \ominus\mathbf{c}]\text{gyr}[\mathbf{c}, \ominus\mathbf{a}](\ominus\mathbf{a}\oplus\mathbf{c}) = \text{gyr}[\mathbf{b}, \ominus\mathbf{c}](\mathbf{c}\ominus\mathbf{a})$$
(3.141)

and, equivalently,

$$\ominus\mathbf{c}\oplus\mathbf{d} = \ominus\text{gyr}[\mathbf{c}, \ominus\mathbf{b}](\ominus\mathbf{b}\oplus\mathbf{a})$$
$$\ominus\mathbf{c}\oplus\mathbf{a} = \ominus\text{gyr}[\mathbf{c}, \ominus\mathbf{b}](\ominus\mathbf{b}\oplus\mathbf{d}) \,.$$
(3.142)

Accordingly, two opposite sides of a gyroparallelogram are congruent, having equal gyrolengths,

$$\|\ominus\mathbf{a}\oplus\mathbf{b}\| = \|\ominus\mathbf{c}\oplus\mathbf{d}\|$$
$$\|\ominus\mathbf{a}\oplus\mathbf{c}\| = \|\ominus\mathbf{b}\oplus\mathbf{d}\| \,.$$
(3.143)

Proof. By Theorem 1.18, p. 16, we have

$$\ominus\mathbf{a}\oplus\mathbf{d} = (\ominus\mathbf{a}\oplus\mathbf{c})\oplus\text{gyr}[\ominus\mathbf{a}, \mathbf{c}](\ominus\mathbf{c}\oplus\mathbf{d})$$
(3.144)

and by Theorem 3.41, p. 93, noting the definition of the gyrogroup cooperation, we have

$$\ominus\mathbf{a}\oplus\mathbf{d} = (\ominus\mathbf{a}\oplus\mathbf{c}) \boxplus (\ominus\mathbf{a}\oplus\mathbf{b}) = (\ominus\mathbf{a}\oplus\mathbf{c})\oplus\text{gyr}[\ominus\mathbf{a}\oplus\mathbf{c}, \mathbf{a}\ominus\mathbf{b}](\ominus\mathbf{a}\oplus\mathbf{b}) \,.$$
(3.145)

Comparing the right-hand side of (3.144) with the extreme right-hand side of (3.145), and employing a left cancellation we have

$$\text{gyr}[\ominus\mathbf{a}\oplus\mathbf{c}, \mathbf{a}\ominus\mathbf{b}](\ominus\mathbf{a}\oplus\mathbf{b}) = \text{gyr}[\ominus\mathbf{a}, \mathbf{c}](\ominus\mathbf{c}\oplus\mathbf{d}) \,.$$
(3.146)

Identity (3.146) can be written in terms of Identity (2.31), p. 38, as

$$\text{gyr}[\mathbf{a}, \ominus\mathbf{c}]\text{gyr}[\mathbf{c}, \ominus\mathbf{b}]\text{gyr}[\mathbf{b}, \ominus\mathbf{a}](\ominus\mathbf{a}\oplus\mathbf{b}) = \text{gyr}[\ominus\mathbf{a}, \mathbf{c}](\ominus\mathbf{c}\oplus\mathbf{d}) \,,$$
(3.147)

which is reducible to the first identity in (3.141) by eliminating $\text{gyr}[\mathbf{a}, \ominus\mathbf{c}]$ on both sides of (3.147). Similarly, interchanging **b** and **c**, one can verify the second identity in (3.141).

The equivalence between (3.141) and (3.142) follows from the gyroautomorphic inverse property and a gyration inversion.

Finally, (3.143) follows from (3.141) since gyrations preserve the gyrolength. □

Theorem 3.43.   (The Gyroparallelogram Gyration Transitive Law). *Let* **abdc** *be a gyroparallelogram in a gyrovector space* $(G, \oplus, \otimes)$. *Then*

$$\text{gyr}[\mathbf{a}, \ominus\mathbf{b}]\text{gyr}[\mathbf{b}, \ominus\mathbf{c}]\text{gyr}[\mathbf{c}, \ominus\mathbf{d}] = \text{gyr}[\mathbf{a}, \ominus\mathbf{d}]. \qquad (3.148)$$

Proof. The proof follows immediately from the gyroparallelogram condition (3.135) and Theorem 2.5, p. 35. □

3.13   POINTS, VECTORS, AND GYROVECTORS

The relationship between points and vectors in the Euclidean geometry of the Euclidean n-space $\mathbb{R}^n$ and the relationship between points and gyrovectors in the hyperbolic geometry of the ball $\mathbb{R}_s^n$ of $\mathbb{R}^n$, $n = 2, 3, \ldots$, involve equivalence relations and equivalence classes of pairs of points.

Definition 3.44.   (Equivalence Relations and Classes). *A relation on a nonempty set S is a subset R of $S \times S$, $R \subset S \times S$, written as $a \sim b$ if $(a, b) \in R$. A relation $\sim$ on a set S is*

(1) *Reflexive if $a \sim a$ for all $a \in S$.*
(2) *Symmetric if $a \sim b$ implies $b \sim a$ for all $a, b \in S$.*
(3) *Transitive if $a \sim b$ and $b \sim c$ imply $a \sim c$ for all $a, b, c \in S$.*

A relation is an equivalence relation if it is reflexive, symmetric, and transitive.

An equivalence relation $\sim$ on a set S gives rise to equivalence classes. The equivalence class of a $\in S$ is the subset $\{x \in S : x \sim a\}$ of S of all the elements $x \in S$ which are related to a by the relation $\sim$.

Two equivalence classes in a set S with an equivalence relation $\sim$ are either equal or disjoint, and the union of all the equivalence classes in S equals S. Accordingly, we say that the equivalence classes of a set S with an equivalence relation form a *partition* of S.

Elements of a real inner product space $\mathbb{V} = (\mathbb{V}, +, \cdot)$, called points and denoted by capital italic letters, A, B, P, Q, etc., give rise to vectors in $\mathbb{V}$, denoted by bold roman lowercase letters **u**, **v**, etc. Any two ordered points $P, Q \in \mathbb{V}$ give rise to a unique rooted vector $\mathbf{v} \in \mathbb{V}$, rooted at the point P. It has a tail at the point P and a head at the point Q, and it has the value $-P + Q$,

$$\mathbf{v} = -P + Q. \qquad (3.149)$$

The length of the rooted vector $\mathbf{v} = -P + Q$ is the distance between the points P and Q, given by the equation

$$\|\mathbf{v}\| = \| - P + Q\|. \qquad (3.150)$$

Two rooted vectors $-P + Q$ and $-R + S$ are equivalent if they have the same value, $-P + Q = -R + S$, that is,

$$- P + Q \sim - R + S \qquad \text{if and only if} \qquad - P + Q = -R + S. \qquad (3.151)$$

The relation $\sim$ in (3.151) between rooted vectors is reflexive, symmetric and transitive. Hence, it is an equivalence relations that gives rise to equivalence classes of rooted vectors. To liberate rooted vectors from their roots we define a *vector* to be an equivalence class of rooted vectors. The vector $-P + Q$ is thus a representative of all rooted vectors with value $-P + Q$.

A point $P \in \mathbb{V}$ is identified with the vector $-O + P$, O being the arbitrarily selected origin of the space $\mathbb{V}$. Hence, the algebra of vectors can be applied to points as well.

Gyrovectors emerge in the ball gyrovector space $\mathbb{V}_s = (\mathbb{V}_s, \oplus, \otimes)$ in a way fully analogous to the way vectors emerge in the space $\mathbb{V}$. Elements of $\mathbb{V}_s$, called points and denoted by capital italic letters, A, B, P, Q, etc., give rise to gyrovectors in $\mathbb{V}_s$, denoted by bold roman lowercase letters $\mathbf{u}$, $\mathbf{v}$, etc. Any two ordered points P, $Q \in \mathbb{V}_s$ give rise to a unique rooted gyrovector $\mathbf{v} \in \mathbb{V}_s$, rooted at the point P. It has a tail at the point P and a head at the point Q, and it has the value $\ominus P \oplus Q$,

$$\mathbf{v} = \ominus P \oplus Q . \tag{3.152}$$

The gyrolength of the rooted gyrovector $\mathbf{v} = \ominus P \oplus Q$ is the gyrodistance between the points P and Q, given by the equation

$$\|\mathbf{v}\| = \|\ominus P \oplus Q\| . \tag{3.153}$$

Two rooted gyrovectors $\ominus P \oplus Q$ and $\ominus R \oplus S$ are equivalent if they have the same value, $\ominus P \oplus Q = \ominus R \oplus S$, that is,

$$\ominus P \oplus Q \quad \sim \quad \ominus R \oplus S \qquad \text{if and only if} \qquad \ominus P \oplus Q = \ominus R \oplus S . \tag{3.154}$$

The relation $\sim$ in (3.154) between rooted gyrovectors is reflexive, symmetric and transitive. Hence, it is an equivalence relations that gives rise to equivalence classes of rooted gyrovectors. To liberate rooted gyrovectors from their roots we define a *gyrovector* to be an equivalence class of rooted gyrovectors. The gyrovector $\ominus P \oplus Q$ is thus a representative of all rooted gyrovectors with value $\ominus P \oplus Q$. Formalizing, we have the following definition of gyrovectors.

Definition 3.45.   (Gyrovectors). *Let $(G, \oplus)$ be a gyrocommutative gyrogroup with its rooted gyrovector equivalence relation $\sim$ in (3.154). The resulting equivalence classes are called gyrovectors. The equivalence class of all rooted gyrovectors that are equivalent to a given rooted gyrovector $PQ = \ominus P \oplus Q$ is the gyrovector denoted by any element of its class.*

A point P of a gyrovector space $(\mathbb{V}_s, \oplus, \otimes)$ is identified with the gyrovector $\ominus O \oplus P$, O being the arbitrarily selected origin of the space $\mathbb{V}_s$. Hence, the algebra of gyrovectors can be applied to points as well.

Following the introduction of our new notation for points and gyrovectors in the paragraph of (3.149) we present Fig. 3.8 of the Einstein gyroparallelogram again, but in the new notation for points in Fig. 3.9, and in the new notation for both points and gyrovectors in Fig. 3.10. Indeed, Fig. 3.9 paves the way for the introduction of the Einstein gyroparallelogram law of gyrovector addition in Fig. 3.10.

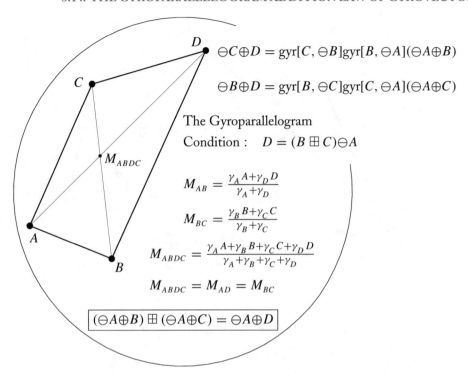

$$\ominus C \oplus D = \text{gyr}[C, \ominus B]\text{gyr}[B, \ominus A](\ominus A \oplus B)$$

$$\ominus B \oplus D = \text{gyr}[B, \ominus C]\text{gyr}[C, \ominus A](\ominus A \oplus C)$$

The Gyroparallelogram

Condition : $D = (B \boxplus C) \ominus A$

$$M_{AB} = \frac{\gamma_A A + \gamma_D D}{\gamma_A + \gamma_D}$$

$$M_{BC} = \frac{\gamma_B B + \gamma_C C}{\gamma_B + \gamma_C}$$

$$M_{ABDC} = \frac{\gamma_A A + \gamma_B B + \gamma_C C + \gamma_D D}{\gamma_A + \gamma_B + \gamma_C + \gamma_D}$$

$$M_{ABDC} = M_{AD} = M_{BC}$$

$$\boxed{(\ominus A \oplus B) \boxplus (\ominus A \oplus C) = \ominus A \oplus D}$$

Figure 3.9: The Einstein gyroparallelogram and its addition law. This figure is identical to Fig. 3.8 with one exception. Here points are denoted by capital italic letters, A, B, C, D, paving the way to Fig. 3.10 where gyrodifferences of points are recognized as gyrovectors, denoted by bold Roman lowercase letters **u**, **v**, **w**.

Consequently, we will find in Sec. 3.14 that gyrovectors are equivalence classes that add according to the gyroparallelogram law, Fig. 3.9, just as vectors are equivalence classes that add according to the parallelogram law.

3.14   THE GYROPARALLELOGRAM ADDITION LAW OF GYROVECTORS

In his 1905 paper that founded the special theory of relativity [9], Einstein noted:

"Das Gesetz vom Parallelogramm der Geschwindigkeiten gilt also nach unserer Theorie nur in erster Annäherung."

A. Einstein [9]

[English translation: Thus the law of velocity parallelogram is valid according to our theory only to a first approximation.]

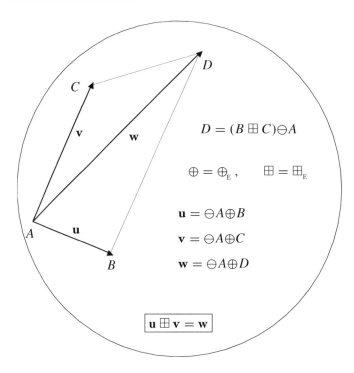

Figure 3.10: The Einstein gyroparallelogram law of gyrovector addition. Let A, B, $C \in \mathbb{R}_s^n$ be any three nongyrocollinear points of an Einstein gyrovector space $(\mathbb{R}_s^n, \oplus_E, \otimes_E)$, giving rise to the two gyrovectors $\mathbf{u} = \ominus A \oplus B$ and $\mathbf{v} = \ominus A \oplus C$, $\oplus = \oplus_E$. Furthermore, let D be a point of the gyrovector space such that $ABDC$ is a gyroparallelogram, that is, $D = (B \boxplus C) \ominus A$ by Def. 3.38 of the gyroparallelogram, $\boxplus = \boxplus_E$. Then, Einstein coaddition of gyrovectors $\mathbf{u}$ and $\mathbf{v}$, $\mathbf{u} \boxplus \mathbf{v} = \mathbf{w}$, expresses the gyroparallelogram law, where $\mathbf{w} = \ominus A \oplus D$. Einstein coaddition, $\boxplus$, thus gives rise to the gyroparallelogram addition law of Einsteinian velocities, which is commutative and fully analogous to the parallelogram addition law of Newtonian velocities. Einsteinian velocities are, thus, gyrovectors that add according to the gyroparalleleogram law just as Newtonian velocities are vectors that add according to the paralleleogram law. Like vectors, a gyrovector $\ominus A \oplus B$ in an Einstein gyrovector space $(\mathbb{R}_s^n, \oplus_E, \otimes_E)$, $n = 2, 3$, is described graphically as a straight arrow from the tail A to the head B.

Fortunately, about a century later it was discovered in [54] that while Einstein velocity addition indeed does not give rise to an exact "velocity parallelogram" in Euclidean geometry, it does give rise to an exact "velocity gyroparallelogram" in hyperbolic geometry, as shown in this section and illustrated in Fig. 3.10 for an Einstein gyrovector plane, and in Fig. 3.11 for a Möbius gyrovector plane.

Let $A, B, C \in \mathbb{R}_s^n$ be any three nongyrocollinear points of an Einstein gyrovector space $(\mathbb{R}_s^n, \oplus, \otimes)$, and let $D \in \mathbb{R}_s^n$ be a point of $\mathbb{R}_s^n$ given by the gyroparallelogram condition, (3.135),

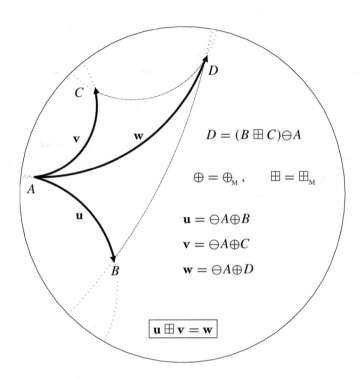

Figure 3.11: The Möbius gyroparallelogram law of gyrovector addition. This figure is similar to Fig. 3.10, except that here $(\mathbb{R}_s^n, \oplus_M, \otimes_M)$ is a Möbius gyrovector plane rather than an Einstein gyrovector plane. Like vectors, a gyrovector $\ominus A \oplus B$, $\oplus = \oplus_M$, in a Möbius gyrovector space $(\mathbb{R}_s^n, \oplus_M, \otimes_M)$, $n = 2, 3$, is described graphically as a circular arrow from the tail A to the head B.

$D = (B \boxplus C)\ominus A$. Then $ABDC$ is a gyroparallelogram in the Einstein gyrovector space $(\mathbb{R}_s^n, \oplus, \otimes)$, obeying the gyroparallelogram law, (3.139),

$$(\ominus A \oplus B) \boxplus (\ominus A \oplus C) = \ominus A \oplus B . \tag{3.155}$$

In Fig. 3.10 we recognize the point gyrodifferences that are shown in Fig. 3.9 and in (3.155) as gyrovectors,

$$\mathbf{u} = \ominus A \oplus B , \qquad \mathbf{v} = \ominus A \oplus C , \qquad \mathbf{w} = \ominus A \oplus D , \tag{3.156}$$

thus obtaining from (3.155) the gyroparallelogram law of gyrovector addition,

$$\mathbf{u} \boxplus \mathbf{v} = \mathbf{w} . \tag{3.157}$$

As shown in Figs. 3.10 and 3.11, the gyrovectors $\mathbf{u}$ and $\mathbf{v}$ in (3.157) generate a gyroparallelogram by forming two adjacent sides of the gyroparallelogram, and the gyrovector $\mathbf{w} = \mathbf{u} \boxplus \mathbf{v}$ in (3.157) forms the resulting gyrodiagonal of the gyroparallelogram.

The gyroparallelogram law (3.157) of gyrovector addition, as shown in Figs. 3.10 and 3.11, is fully analogous to the parallelogram law of vector addition in Euclidean geometry, and is given by the coaddition law of gyrovectors. Remarkably, we see here once again that in order to capture analogies with classical results we must employ both the gyrocommutative operation $\oplus$ and the commutative cooperation $\boxplus$ of gyrovector spaces.

The breakdown of commutativity in Einstein velocity addition law seemed undesirable to the famous mathematician Emile Borel. Borel's resulting attempt to "repair" the seemingly "defective" Einstein velocity addition in the years following 1912 is described by Walter in [64, p. 117]. Here, however, we discover that there is no need to repair Einstein velocity addition law for being non-commutative since, despite of being noncommutative, it gives rise to the gyroparallelogram law of gyrovector addition, which turns out to be commutative. The compatibility of our gyroparallelogram addition law of Einsteinian velocities in Fig. 3.10 with cosmological observations of stellar aberration will be discussed in Sec. 4.23. It is demonstrated in Sec. 4.23 that the well-known relativistic stellar aberration formulas, which are obtained in the literature by employing the Lorentz transformation group, can be recovered by applying the hyperbolic geometry and trigonometry of an Einstein gyrovector space.

3.15   GYROVECTOR GYROTRANSLATION

It is often necessary to move a rooted gyrovector without distorting its value, so that its tail lines up with that of another rooted gyrovector. Once two rooted gyrovectors possess a common tail:

(i) they form a gyroangle, as shown in Fig. 4.2, p. 115, for Möbius gyrovectors; and

(ii) they can be added by the gyroparallelogram law, as shown in Fig. 3.10 for Einstein gyrovectors, and in Fig. 3.11 for Möbius gyrovectors.

The motion of a rooted gyrovector in its gyrovector space without distorting its value is called a *gyrovector gyrotranslation*. A formal definition of the latter, based on Theorem 3.46 below, is given in Def. 3.47.

A graphical illustration of a gyrovector gyrotranslation of a gyrovector $\ominus A \oplus B$ into another gyrovector $\ominus A' \oplus B'$ is presented in Fig. 3.13, p. 104, along with its vector space counterpart, in Fig. 3.12, p. 104.

The following Theorem 3.46 and Definition 3.47, accordingly, allow rooted gyrovector equivalence to be expressed in terms of rooted gyrovector gyrotranslation.

Theorem 3.46.   *Let*

$$PQ = \ominus P \oplus Q$$
$$P'Q' = \ominus P' \oplus Q'$$

(3.158)

be two rooted gyrovectors in a gyrocommutative gyrogroup $(G, \oplus)$. *They are equivalent, that is,*

$$\ominus P \oplus Q = \ominus P' \oplus Q', \tag{3.159}$$

if and only if there exists a gyrovector $\mathbf{t} \in G$ *such that*

$$\begin{aligned} P' &= \mathrm{gyr}[P, \mathbf{t}](\mathbf{t} \oplus P) \\ Q' &= \mathrm{gyr}[P, \mathbf{t}](\mathbf{t} \oplus Q) . \end{aligned} \tag{3.160}$$

Furthermore, the gyrovector $\mathbf{t}$ *is unique, given by the equation*

$$\mathbf{t} = \ominus P \oplus P' . \tag{3.161}$$

Proof. By the Gyrotranslation Theorem 2.11, p. 37, we have

$$\ominus(\mathbf{t} \oplus P) \oplus (\mathbf{t} \oplus Q) = \mathrm{gyr}[\mathbf{t}, P](\ominus P \oplus Q) . \tag{3.162}$$

Hence, by the gyroautomorphism inversion law, we have

$$\ominus P \oplus Q = \mathrm{gyr}[P, \mathbf{t}]\{\ominus(\mathbf{t} \oplus P) \oplus (\mathbf{t} \oplus Q)\} \tag{3.163}$$

for any $P, Q, \mathbf{t} \in G$.

Assuming (3.160), we have by (3.163) and (3.160)

$$\begin{aligned} \ominus P \oplus Q &= \mathrm{gyr}[P, \mathbf{t}]\{\ominus(\mathbf{t} \oplus P) \oplus (\mathbf{t} \oplus Q)\} \\ &= \ominus \mathrm{gyr}[P, \mathbf{t}](\mathbf{t} \oplus P) \oplus \mathrm{gyr}[P, \mathbf{t}](\mathbf{t} \oplus Q) \\ &= \ominus P' \oplus Q', \end{aligned} \tag{3.164}$$

thus verifying (3.159).

Conversely, assuming (3.159) we let

$$\mathbf{t} = \ominus P \oplus P', \tag{3.165}$$

so that by a left cancellation and by the gyrocommutative law we have

$$P' = P \oplus \mathbf{t} = \mathrm{gyr}[P, \mathbf{t}](\mathbf{t} \oplus P) , \tag{3.166}$$

thus verifying the first equation in (3.160).

We will now verify that (3.159) implies the second equation in (3.160) as well. With the notation $g_{P,\mathbf{t}} = \mathrm{gyr}[P, \mathbf{t}]$ whenever convenient we have the following chain of equalities, which are numbered for subsequent derivation:

$$
\begin{aligned}
Q' &\overset{(1)}{=\!=\!=} P'\oplus(\ominus P\oplus Q) \\
&\overset{(2)}{=\!=\!=} \mathrm{gyr}[P,\mathbf{t}](\mathbf{t}\oplus P)\oplus(\ominus P\oplus Q) \\
&\overset{(3)}{=\!=\!=} (g_{P,\mathbf{t}}\mathbf{t}\oplus g_{P,\mathbf{t}}P)\oplus(\ominus P\oplus Q) \\
&\overset{(4)}{=\!=\!=} g_{P,\mathbf{t}}\mathbf{t}\oplus\{g_{P,\mathbf{t}}P\oplus\mathrm{gyr}[g_{P,\mathbf{t}}P,\,g_{P,\mathbf{t}}\mathbf{t}](\ominus P\oplus Q)\} \\
&\overset{(5)}{=\!=\!=} g_{P,\mathbf{t}}\mathbf{t}\oplus\{g_{P,\mathbf{t}}P\oplus g_{P,\mathbf{t}}(\ominus P\oplus Q)\} \\
&\overset{(6)}{=\!=\!=} g_{P,\mathbf{t}}\mathbf{t}\oplus\{g_{P,\mathbf{t}}P\oplus(\ominus g_{P,\mathbf{t}}P\oplus g_{P,\mathbf{t}}Q)\} \\
&\overset{(7)}{=\!=\!=} g_{P,\mathbf{t}}\mathbf{t}\oplus g_{P,\mathbf{t}}Q \\
&\overset{(8)}{=\!=\!=} g_{P,\mathbf{t}}(\mathbf{t}\oplus Q),
\end{aligned}
\tag{3.167}
$$

thus verifying the second equation in (3.160), as desired.

Derivation of the numbered equalities in (3.167) follows.

(1) Follows from (3.159) by a left cancellation.

(2) Follows from (1) by a substitution of P' from the first equation in (3.160).

(3) Follows from (2) by applying the gyration $g_{P,\mathbf{t}} = \mathrm{gyr}[P,\mathbf{t}]$ to $(\mathbf{t}\oplus P)$ term by term, noting that gyrations respect the gyrogroup operation.

(4) Follows from (3) by applying the right gyroassociative law.

(5) Follows from (4) by Identity (1.77), p. 20.

(6) Follows from (5) by applying the gyration $g_{P,\mathbf{t}} = \mathrm{gyr}[P,\mathbf{t}]$ to $(\ominus P\oplus Q)$ term by term.

(7) Follows from (6) by a left cancellation.

(8) Follows from (7) immediately, noting that gyrations respect the gyrogroup operation.

Finally, the gyrovector $\mathbf{t}\in G$ is uniquely determined by P and P' as we see from the first equation in (3.160) and the gyrocommutative law,

$$
\ominus P\oplus P' = \ominus P\oplus\mathrm{gyr}[P,\mathbf{t}](\mathbf{t}\oplus P) = \ominus P\oplus(P\oplus\mathbf{t}) = \mathbf{t}.
\tag{3.168}
$$

□

Theorem 3.46 characterizes rooted gyrovector equivalence in terms of a gyrotranslation by a gyrovector followed by a gyration, (3.160). As such, it suggests the following definition of *gyrovector gyrotranslation* in gyrocommutative gyrogroups.

Definition 3.47.   (Gyrovector Gyrotranslation). *A gyrovector gyrotranslation $T_{\mathbf{t}}$ by a gyrovector $\mathbf{t}\in G$ of a rooted gyrovector $PQ = \ominus P\oplus Q$, with tail P and head Q, in a gyrocommutative gyrogroup*

$(G, \oplus)$ *is the rooted gyrovector* $T_{\mathbf{t}} P Q = P'Q' = \ominus P' \oplus Q'$, *with tail* P' *and head* Q' *given by* (3.160), *that is,*

$$P' = \mathrm{gyr}[P, \mathbf{t}](\mathbf{t} \oplus P)$$
$$Q' = \mathrm{gyr}[P, \mathbf{t}](\mathbf{t} \oplus Q) \,. \tag{3.169}$$

The rooted gyrovector $\ominus P' \oplus Q'$ *is said to be the* $\mathbf{t}$ *gyrovector gyrotranslation, or the gyrovector gyrotranslation by* $\mathbf{t}$*, of the rooted gyrovector* $\ominus P \oplus Q$.

We may note that the two equations in (3.169) are not symmetric in P and Q since they share a gyration that depends on P but is independent of Q. Moreover, owing to the gyrocommutativity of $\oplus$, the first equation in (3.169) can be written as

$$P' = P \oplus \mathbf{t} \,. \tag{3.170}$$

As expected, a gyrovector gyrotranslation by the zero gyrovector $\mathbf{0} \in G$ is trivial, since (3.169) reduces to

$$P' = \mathrm{gyr}[P, \mathbf{0}](\mathbf{0} \oplus P) = P$$
$$Q' = \mathrm{gyr}[P, \mathbf{0}](\mathbf{0} \oplus Q) = Q \,. \tag{3.171}$$

Definition 3.47 allows Theorem 3.46 to be reformulated as follows.

Theorem 3.48.   *Two rooted gyrovectors*

$$P Q = \ominus P \oplus Q$$
$$P' Q' = \ominus P' \oplus Q' \tag{3.172}$$

in a gyrocommutative gyrogroup $(G, \oplus)$ *are equivalent, that is,*

$$\ominus P \oplus Q = \ominus P' \oplus Q' \,, \tag{3.173}$$

if and only if gyrovector $P' Q'$ *is a gyrovector gyrotranslation of gyrovector* $P Q$. *Furthermore, if* $P' Q'$ *is a gyrovector gyrotranslation of* $P Q$ *then it is a gyrovector gyrotranslation of* $P Q$ *by*

$$\mathbf{t} = \ominus P \oplus P' \,. \tag{3.174}$$

Analogies that the familiar vector translation in vector spaces and their Euclidean geometry shares with gyrovector gyrotranslation in gyrovector spaces and their hyperbolic geometry are illustrated in Figs. 3.12 and 3.13.

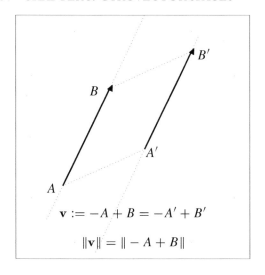

Figure 3.12: The vector $-A' + B'$ is a translation of the vector $-A + B$ by a vector $\mathbf{t}$ in a Euclidean plane $\mathbb{R}^2$, given by (3.175). As such, these two vectors are equivalent and, hence, indistinguishable in their vector space and its underlying Euclidean geometry. Unlike hyperbolic geometry, where parallelism is denied, two equivalent nonzero vectors in Euclidean geometry are parallel, as shown here. The quadrilateral $AA'B'B$ is a parallelogram, presenting a disanalogy with Fig. 3.13.

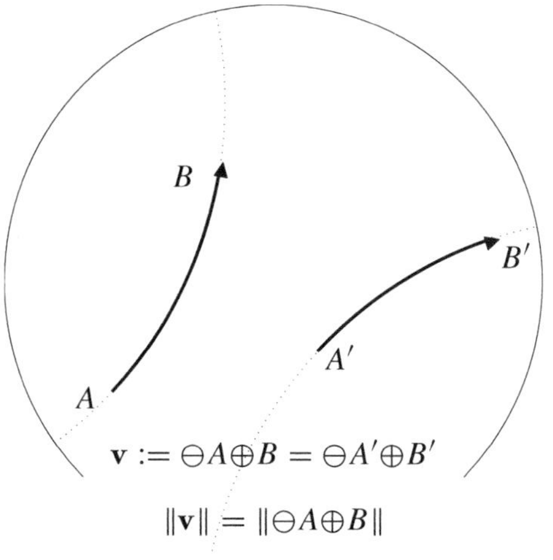

Figure 3.13: The gyrovector $\ominus A' \oplus B'$ is a gyrovector gyrotranslation of the gyrovector $\ominus A \oplus B$ by a gyrovector $\mathbf{t}$ in a Möbius gyrovector plane $(\mathbb{R}^2_s, \oplus, \otimes)$, given by (3.169). The analogies that (3.169) shares with its Euclidean counterpart (3.175) are obvious. As such, these two gyrovectors are equivalent and, hence, indistinguishable in their gyrovector space and its underlying hyperbolic geometry. The gyroquadrilateral $AA'B'B$ is not a gyroparallelogram.

In Fig. 3.12, A and B are two distinct points in the Euclidean plane $\mathbb{R}^2$. The resulting vector $-A + B$ is $\mathbf{t}$ translated (that is, translated by the vector $\mathbf{t}$) into the vector $-A' + B'$ in the Euclidean plane $\mathbb{R}^2$ according to the vector translation formula

$$\begin{aligned} A' &= \mathbf{t} + A \\ B' &= \mathbf{t} + B \,. \end{aligned} \tag{3.175}$$

It follows from (3.175) immediately that the original vector $-A + B$ and the translated vector $-A' + B'$ in Fig. 3.12 have equal values,

$$- A' + B' = -(\mathbf{t} + A) + (\mathbf{t} + B) = -A + B \,, \tag{3.176}$$

and, hence, they are equivalent. As vectors, rather than rooted vectors, the two vectors $-A + B$ and $-A' + B'$ in Fig. 3.12 are thus indistinguishable.

A gyro-analog of vector translation in the Euclidean plane $\mathbb{R}^2$ in Fig. 3.12 is the gyrovector gyrotranslation in the Poincaré disc model of hyperbolic geometry, shown in Fig. 3.13. In this figure A and B are two distinct points in a Möbius gyrovector plane $\mathbb{R}_s^2$. The resulting gyrovector $\ominus A \oplus B$ is $\mathbf{t}$ gyrotranslated (that is, gyrotranslated by the gyrovector $\mathbf{t}$) into the gyrovector $\ominus A' \oplus B'$ in the Möbius gyrovector plane $(\mathbb{R}_s^2, \oplus, \otimes)$ according to the gyrovector gyrotranslation formula (3.169). This gyrovector gyrotranslation formula is the gyro-analog of the Euclidean vector translation formula (3.175).

It follows from the gyrovector gyrotranslation formula (3.169), by Theorem 3.48, that the original gyrovector $\ominus A \oplus B$ and its gyrotranslated gyrovector $\ominus A' \oplus B'$ in Fig. 3.13 have equal values,

$$\ominus A \oplus B = \ominus A' \oplus B', \tag{3.177}$$

and, hence, they are equivalent. As gyrovectors, rather than rooted gyrovectors, the two gyrovectors $\ominus A \oplus B$ and $\ominus A' \oplus B'$ in Fig. 3.13 are thus indistinguishable.

Along with analogies, vector translation and gyrovector gyrotranslation have an important disanalogy. A vector and its translated vector by a vector $\mathbf{t} \neq \mathbf{0}$ form a parallelogram as, for instance, the parallelogram $AA'B'B$ in Fig. 3.12. In contrast, a gyrovector and its gyrotranslated gyrovector by a gyrovector $\mathbf{t} \neq \mathbf{0}$ do not form a gyroparallelogram. Thus, for instance, the gyroquadrilateral $AA'B'B$ in Fig. 3.13 is not a gyroparallelogram.

The following theorem enables us to move gyrovectors to any given points of their gyrovector space.

Theorem 3.49.   (Gyrovector Gyrotranslation of Head and Tail of Gyrovectors). *Let P, Q, P', and Q' be any four points of a gyrocommutative gyrogroup $(G, \oplus)$.*

(1) *The gyrovector gyrotranslation of the rooted gyrovector $PQ = \ominus P \oplus Q$ into the rooted gyrovector $P'X = \ominus P' \oplus X$ with a given tail P', determines the head X of $P'X$,*

$$X = P' \oplus (\ominus P \oplus Q), \tag{3.178}$$

so that

$$P'X = \ominus P' \oplus \{P' \oplus (\ominus P \oplus Q)\}. \tag{3.179}$$

(2) *The gyrovector gyrotranslation of the rooted gyrovector $PQ = \ominus P \oplus Q$ into the rooted gyrovector $YQ' = \ominus Y \oplus Q'$, with a given head Q', determines the tail Y of YQ',*

$$Y = (P \ominus Q) \boxplus Q', \tag{3.180}$$

so that

$$YQ' = \ominus \{(P \ominus Q) \boxplus Q'\} \oplus Q'. \tag{3.181}$$

Proof. (1) The gyrovector $P'X$ is a gyrovector gyrotranslation of PQ. Hence, by Theorem 3.48, PQ and $P'X$ are equivalent gyrovectors. Hence, by (3.154) we have,

$$\ominus P \oplus Q = \ominus P' \oplus X \,, \tag{3.182}$$

from which (3.178) follows by a left cancellation.

(2) The gyrovector YQ' is a gyrovector gyrotranslation of PQ. Hence, by Theorem 3.48, PQ and YQ' are equivalent gyrovectors. Hence, by (3.154) we have,

$$\ominus P \oplus Q = \ominus Y \oplus Q' \,, \tag{3.183}$$

from which (3.180) follows by a right cancellation, (1.72), p. 19, along with the gyroautomorphic inverse property in Theorem 2.2, p. 33, and the cogyroautomorphic inverse property in Theorem 1.32, p. 29. □

3.16   GYROVECTOR GYROTRANSLATION COMPOSITION

Two successive gyrovector gyrotranslations are equivalent to a single gyrovector gyrotranslation. Owing to its importance, the resulting composite gyrovector gyrotranslation is explored in this section.

Let $P''Q''$ be the gyrovector gyrotranslation of a rooted gyrovector $P'Q'$ by $\mathbf{t}_2$ where $P'Q'$, in turn, is the gyrovector gyrotranslation of a rooted gyrovector PQ by $\mathbf{t}_1$ in a gyrocommutative gyrogroup $(G, \oplus)$. Then, by Def. 3.47,

$$\begin{aligned} P'' &= \operatorname{gyr}[P', \mathbf{t}_2](\mathbf{t}_2 \oplus P') = P' \oplus \mathbf{t}_2 \\ P' &= \operatorname{gyr}[P, \mathbf{t}_1](\mathbf{t}_1 \oplus P) = P \oplus \mathbf{t}_1 \,, \end{aligned} \tag{3.184}$$

so that

$$\begin{aligned} P'' &= P' \oplus \mathbf{t}_2 \\ &= (P \oplus \mathbf{t}_1) \oplus \mathbf{t}_2 \\ &= P \oplus (\mathbf{t}_1 \oplus \operatorname{gyr}[\mathbf{t}_1, P]\mathbf{t}_2) \,. \end{aligned} \tag{3.185}$$

Moreover, by Theorem 3.48 the rooted gyrovector $P''Q''$ is equivalent to the rooted gyrovector $P'Q'$ and the latter, in turn, is equivalent to the rooted gyrovector PQ. Hence, $P''Q''$ is equivalent to PQ, so that, by Theorem 3.46, $P''Q''$ is a gyrovector gyrotranslation of PQ by some unique $\mathbf{t}_{12} \in G$,

$$P'' = P \oplus \mathbf{t}_{12} \,. \tag{3.186}$$

Comparing (3.186) and (3.185), we see that $\mathbf{t}_{12}$ is given by the equation

$$\mathbf{t}_{12} = \mathbf{t}_1 \oplus \operatorname{gyr}[\mathbf{t}_1, P]\mathbf{t}_2 \,. \tag{3.187}$$

Expressing the rooted gyrovector $P'Q'$ in terms of the rooted gyrovector PQ we have, by Theorem 3.46 and the gyrocommutative law,

$$P' = P \oplus \mathbf{t}_1$$
$$Q' = \text{gyr}[P, \mathbf{t}_1](\mathbf{t}_1 \oplus Q) . \tag{3.188}$$

Similarly, expressing the rooted gyrovector $P''Q''$ in terms of the rooted gyrovector $P'Q'$ we have, by Theorem 3.46 and the gyrocommutative law,

$$P'' = P' \oplus \mathbf{t}_2$$
$$Q'' = \text{gyr}[P', \mathbf{t}_2](\mathbf{t}_2 \oplus Q') . \tag{3.189}$$

Finally, expressing the rooted gyrovector $P''Q''$ in terms of the rooted gyrovector PQ we have, by Theorem 3.46 and the gyrocommutative law,

$$P'' = P \oplus \mathbf{t}_{12}$$
$$Q'' = \text{gyr}[P, \mathbf{t}_{12}](\mathbf{t}_{12} \oplus Q) . \tag{3.190}$$

Substituting (3.187) into (3.190) we have

$$P'' = P \oplus (\mathbf{t}_1 \oplus \text{gyr}[\mathbf{t}_1, P]\mathbf{t}_2)$$
$$Q'' = \text{gyr}[P, \mathbf{t}_1 \oplus \text{gyr}[\mathbf{t}_1, P]\mathbf{t}_2]\{(\mathbf{t}_1 \oplus \text{gyr}[\mathbf{t}_1, P]\mathbf{t}_2) \oplus Q\} . \tag{3.191}$$

Substituting (3.188) into the second equation in (3.189) we have

$$Q'' = \text{gyr}[P \oplus \mathbf{t}_1, \mathbf{t}_2]\{\mathbf{t}_2 \oplus \text{gyr}[P, \mathbf{t}_1](\mathbf{t}_1 \oplus Q)\} . \tag{3.192}$$

From (3.192) and the second equation in (3.191) for Q'' we have the identity

$$\text{gyr}[P, \mathbf{t}_1 \oplus \text{gyr}[\mathbf{t}_1, P]\mathbf{t}_2]\{(\mathbf{t}_1 \oplus \text{gyr}[\mathbf{t}_1, P]\mathbf{t}_2) \oplus Q\}$$
$$= \text{gyr}[P \oplus \mathbf{t}_1, \mathbf{t}_2]\{\mathbf{t}_2 \oplus \text{gyr}[P, \mathbf{t}_1](\mathbf{t}_1 \oplus Q)\} \tag{3.193}$$

for all $P, Q, \mathbf{t}_1, \mathbf{t}_2 \in G$.

Thus, in our way to uncover the composition law (3.190) of gyrovector gyrotranslation we obtained the new gyrocommutative gyrogroup identity (3.193) as an unintended and unforeseen by-product. Interestingly, the new identity (3.193) reduces to (2.23), p. 36, when $P = O$.

The composite gyrovector gyrotranslation (3.190) is trivial when $\mathbf{t}_{12} = \mathbf{0}$, that is, when $\mathbf{t}_2 = \ominus \text{gyr}[P, \mathbf{t}_1]\mathbf{t}_1$, as we see from (3.187). Hence, the inverse gyrovector gyrotranslation of a gyrovector gyrotranslation by $\mathbf{t}$ of a rooted gyrovector PQ is a gyrovector gyrotranslation by $\ominus \text{gyr}[P, \mathbf{t}]\mathbf{t}$.

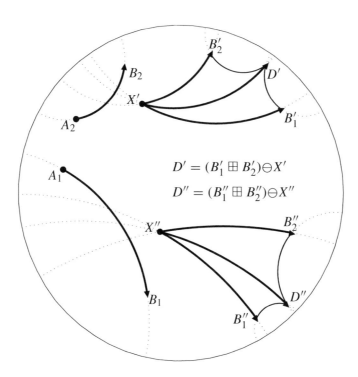

Figure 3.14: The Möbius gyroparallelogram law of gyrovector addition in a Möbius gyrovector plane. Two gyrovectors, $\ominus A_1 \oplus B_1$ and $\ominus A_2 \oplus B_2$, are gyrovector gyrotranslated to a common tail where they generate a gyroparallelogram, like the two gyroparallelograms shown here with tails X' and X''. The resulting gyrovector sums $\ominus X' \oplus D'$ and $\ominus X'' \oplus D''$ are equivalent and, hence, indistinguishable.

3.17   GYROVECTOR GYROTRANSLATION AND THE GYROPARALLELOGRAM LAW

In order to capture analogies with the Euclidean geometry of vector spaces, vector translations in Euclidean geometry go over to two kinds of gyrovector translations. These are left gyrotranslations of gyrovectors and gyrovector gyrotranslations, where the former is governed by Theorem 2.11, p. 37, and the latter, studied in Sec. 3.15, is what we need for the application of the gyroparallelogram addition law of gyrovectors.

The gyroparallelogram law of gyrovector addition is applied in Fig. 3.10 to two gyrovectors that possess a common tail. Owing to the gyrovector gyrotranslation technique presented in sec. 3.16, one may apply the gyroparallelogram law to add two gyrovectors with arbitrary tails by gyrovector gyrotranslating the two gyrovectors to an arbitrarily selected common tail in their gyrovector space. A detailed illustration for Möbius gyrovectors and the Möbius gyroparallelogram law of gyrovector addition in a Möbius gyrovector plane is presented in Fig. 3.14.

In Fig. 3.14 we consider in a Möbius gyrovector plane $(\mathbb{R}_s^2, \oplus, \otimes)$ two given gyrovectors, $\ominus A_1 \oplus B_1$ and $\ominus A_2 \oplus B_2$, that need not have a common tail, $A_1 \neq A_2$. To visualize the gyroparallelogram addition of the two given gyrovectors, the gyrovectors must be gyrovector gyrotranslated to an arbitrarily selected common tail, where they form a gyroparallelogram. Thus, for instance, (i) at the tail X' they form the gyroparallelogram $X'B_1'D'B_2'$ with the gyrodiagonal gyrovector $\ominus X' \oplus D'$, and (ii) at the tail X'' they form the gyroparallelogram $X''B_1''D''B_2''$ with the gyrodiagonal gyrovector $\ominus X'' \oplus D''$.

The gyrodiagonal of each resulting gyroparallelogram forms a gyrovector that gives the sum of the two given gyrovectors. Naturally, in full analogy with Euclidean geometry, the resulting gyrovector sum is independent of the choice of the common tail. For the two distinct selections, X' and X'', of the common tail in Fig. 3.14 we thus have the identity

$$(\ominus A_1 \oplus B_1) \boxplus (\ominus A_2 \oplus B_2) = \ominus X' \oplus D' = \ominus X'' \oplus D''. \tag{3.194}$$

3.18 THE MÖBIUS GYROTRIANGLE GYROANGLES

Following the graphical presentation of

(1) gyrolines, in Figs. 3.1–3.2 and Figs. 3.3–3.4;

(2) gyrovectors, in Fig. 3.13, and in Figs. 3.10–3.11;

(3) gyrotriangles, in Figs. 3.6–3.7; and

(4) gyroparallelograms, in Figs. 3.10–3.11,

in both Einstein and Möbius gyrovector planes, we present here, in Fig. 3.15, the Möbius gyrotriangle along with its standard notation in a Möbius gyrovector space $(\mathbb{V}_s, \oplus, \otimes)$. An Einstein gyrotriangle, along with the same standard notation, is presented in Fig. 4.4, p. 117.

The Möbius gyrotriangle is studied in Sec. 3.11. A Möbius gyrotriangle ABC, Fig. 3.15, is formed by any three non-gyrocollinear points A, B, and C, which are the gyrotriangle vertices. The three sides of gyrotriangle ABC are the gyrosegments AB, BC, and CA. These gyrotriangle gyrosegments form the gyrovectors

$$\begin{aligned}
\mathbf{a} &= \ominus B \oplus C \\
\mathbf{b} &= \ominus C \oplus A \\
\mathbf{c} &= \ominus A \oplus B.
\end{aligned} \tag{3.195}$$

The gyrolengths a, b, and c of the gyrovectors $\mathbf{a}$, $\mathbf{b}$, and $\mathbf{c}$, shown in Fig. 3.15, are the gyrotriangle side gyrolengths. In addition, gyrotriangle ABC has three gyroangles, α, β, and γ. These gyroangles lead to *gyrotrigonometry*, the study of how the side gyrolengths and gyroangle measures of a gyrotriangle are related to each other.

Gyrotrigonometry is studied in Chap. 4 in a way fully analogous to the common study of trigonometry, as indicated in Fig. 3.15.

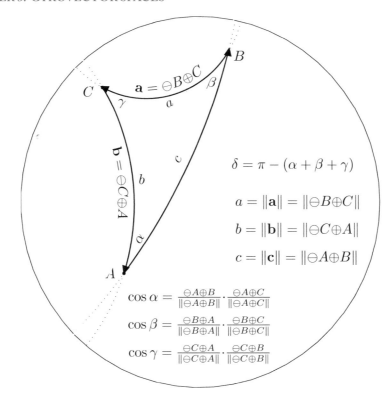

Figure 3.15: The gyrotriangle ABC in a Möbius gyrovector space $(\mathbb{V}_s, \oplus, \otimes)$ is presented along with its associated standard notation. The gyrotriangle vertices, A, B, C, are any non-gyrocollinear points of $\mathbb{V}_s$. For the hyperbolic plane, $\mathbb{V}_s = \mathbb{R}_s^2$, the gyrotriangle sides are presented graphically as gyroline gyrosegments that join the vertices. The gyrotriangle sides form gyrovectors, $\mathbf{a}, \mathbf{b}, \mathbf{c}$, side-gyrolengths, a, b, c, and gyroangles, α, β, γ. The gyrotriangle gyroangle sum is less than π the difference, $\delta = \pi - (\alpha_1 + \alpha_2 + \alpha_3)$, being the gyrotriangular defect. The gyrocosine function of the gyrotriangle gyroangles is presented. Remarkably, it assumes a form that is fully analogous to the cosine function of Euclidean trigonometry, as we will see in Chap. 4.

3.19   EXERCISES

(1) Prove Identity (3.72), p. 72. Hint: see [61, Sec. 6.6].

(2) Prove Identities (3.106) – (3.111), p. 81.

(3) Use linear algebra to prove Identity (3.119), p. 85, of Theorem (3.35), p. 85, in detail.

(4) Prove that, as expected in Def. 3.32, p. 81, the triple $(\mathbb{V}_s, \oplus_{\text{E}}, \otimes_{\text{E}})$ is a gyrovector space.

CHAPTER 4

Gyrotrigonometry

Gyrotrigonometry is the study of how the sides and gyroangles of a gyrotriangle are related to each other.

4.1 THE GYROANGLE

Definition 4.1. (Unit Gyrovectors). *Let A and B be two distinct points, and let $\ominus A \oplus B$ be the resulting nonzero gyrovector in a gyrovector space $(G, \oplus, \otimes)$. Its gyrolength is $\|\ominus A \oplus B\| \neq 0$, and its associated gyrovector*

$$\frac{\ominus A \oplus B}{\|\ominus A \oplus B\|} \tag{4.1}$$

is called a unit gyrovector.

Guided by analogies with Euclidean trigonometry, the definition of the gyrocosine of a gyroangle follows.

Definition 4.2. (The Gyrocosine Function and Gyroangles, I). *Let $\ominus A \oplus B$ and $\ominus A \oplus C$ be two nonzero rooted gyrovectors rooted at a common point A in a gyrovector space $(G, \oplus, \otimes)$. The gyrocosine of the measure of the gyroangle α, $0 \leq \alpha \leq \pi$, that the two rooted gyrovectors include is given by the equation*

$$\cos \alpha = \frac{\ominus A \oplus B}{\|\ominus A \oplus B\|} \cdot \frac{\ominus A \oplus C}{\|\ominus A \oplus C\|} . \tag{4.2}$$

The gyroangle α in (4.2) is denoted by $\alpha = \angle BAC$ or, equivalently, $\alpha = \angle CAB$. Two gyroangles are congruent if they have the same measure.

Gyroangles in a Möbius and in an Einstein gyrovector plane are shown in Figs. 4.1 and 4.4, respectively.

We will soon find that the functional properties of the gyrocosine function, cos, of gyrotrigonometry in (4.2) are identical with those of the cosine function of trigonometry. Hence, the use of the same notation, cos, for both the gyrocosine function of gyrotrigonometry in Figs. 4.1 – 4.4 and the common cosine function of trigonometry will prove justified and useful.

In Def. 4.5, p. 115, the gyroangle definition in Def. 4.2, shown in Fig. 4.1, will be extended to a gyroangle that is determined by two nonzero gyrovectors that need not be rooted at the same point.

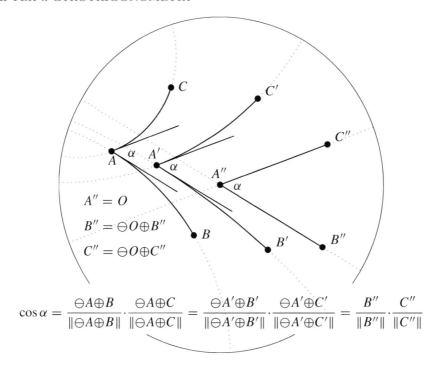

$$\cos\alpha = \frac{\ominus A \oplus B}{\|\ominus A \oplus B\|} \cdot \frac{\ominus A \oplus C}{\|\ominus A \oplus C\|} = \frac{\ominus A' \oplus B'}{\|\ominus A' \oplus B'\|} \cdot \frac{\ominus A' \oplus C'}{\|\ominus A' \oplus C'\|} = \frac{B''}{\|B''\|} \cdot \frac{C''}{\|C''\|}$$

Figure 4.1: A Möbius gyroangle α formed by two intersecting Möbius geodesic rays (gyrorays) in a Möbius gyrovector space $(\mathbb{R}^n_s, \oplus, \otimes)$. Its measure equals the measure of the Euclidean angle formed by corresponding intersecting tangent lines. As such, the Möbius gyroangle is *additive*, that is, if a Möbius gyroangle is split into two gyroangles then its measure equals the sum of the measures of these gyroangles. Gyroangles are invariant under left gyrotranslations, (4.6). Shown are two successive left gyrotranslations of the gyroangle $\alpha = \angle BAC$ into $\alpha = \angle B'A'C'$, and the latter into $\alpha = \angle B''A''C'' = \angle B''OC''$, so that $\cos\alpha = (B''/\|B''\|) \cdot (C''/\|C''\|)$, as in Euclidean geometry.

Theorem 4.3. *Gyroangles are invariant under gyrovector space automorphisms and left gyrotranslations.*

Proof. Let A, B, C be any three distinct points of a gyrovector space $(G, \oplus, \otimes)$. Their images under any automorphism $\tau \in Aut(G, \oplus, \otimes)$, Def. 3.7, p. 60, are $\tau A, \tau B, \tau C$, and their left gyrotranslations by any $X \in G$ are $X \oplus A$, $X \oplus B$, $X \oplus C$, respectively. We have to prove that $\angle ABC$ is invariant under gyrovector space automorphisms, that is,

$$\angle BAC = \angle(\tau B)(\tau A)(\tau C), \tag{4.3}$$

and that $\angle ABC$ is invariant under left gyrotranslations, that is,

$$\angle BAC = \angle(X \oplus B)(X \oplus A)(X \oplus C), \tag{4.4}$$

for all A, B, C, $X \in G$ and all $\tau \in Aut(G, \oplus, \otimes)$, Fig. 4.1.

Employing (3.18), p. 60, we have by Def. 4.2

$$\cos\angle(\tau B)(\tau A)(\tau C) = \frac{\ominus\tau A\oplus\tau B}{\|\ominus\tau A\oplus\tau B\|}\cdot\frac{\ominus\tau A\oplus\tau C}{\|\ominus\tau A\oplus\tau C\|}$$

$$= \frac{\tau(\ominus A\oplus B)}{\|\tau(\ominus A\oplus B)\|}\cdot\frac{\tau(\ominus A\oplus C)}{\|\tau(\ominus A\oplus C)\|} \qquad (4.5)$$

$$= \frac{\ominus A\oplus B}{\|\ominus A\oplus B\|}\cdot\frac{\ominus A\oplus C}{\|\ominus A\oplus C\|}$$

$$= \cos\angle BAC\,,$$

since automorphisms preserve the inner product and the norm, thus verifying (4.3).

In order to verify (4.4), let us consider the following chain of equations, which are numbered for subsequent explanation:

$$\cos\angle(X\oplus B)(X\oplus A)(X\oplus C) \overset{(1)}{=\!=} \frac{\ominus(X\oplus A)\oplus(X\oplus B)}{\|\ominus(X\oplus A)\oplus(X\oplus B)\|}\cdot\frac{\ominus(X\oplus A)\oplus(X\oplus C)}{\|\ominus(X\oplus A)\oplus(X\oplus C)\|}$$

$$\overset{(2)}{=\!=} \frac{\mathrm{gyr}[X,A](\ominus A\oplus B)}{\|\mathrm{gyr}[X,A](\ominus A\oplus B)\|}\cdot\frac{\mathrm{gyr}[X,A](\ominus A\oplus C)}{\|\mathrm{gyr}[X,A](\ominus A\oplus C)\|} \qquad (4.6)$$

$$\overset{(3)}{=\!=} \frac{\ominus A\oplus B}{\|\ominus A\oplus B\|}\cdot\frac{\ominus A\oplus C}{\|\ominus A\oplus C\|}$$

$$\overset{(4)}{=\!=} \cos\angle BAC\,,$$

thus verifying (4.4).

Derivation of the numbered equalities in (4.6) follows.

(1) Follows from the gyroangle definition in Def. 4.2.

(2) Follows from (1) by the Gyrotranslation Theorem 2.11, p. 37. It is useful to note that the gyrations in (2), $\mathrm{gyr}[X,A]$, are independent of B and C, as remarked in Remark 2.12, p. 38.

(3) Follows from (2) by Axiom $(V1)$ of gyrovector spaces, according to which gyrations preserve the inner product and, hence, the norm.

(4) Follows from (3) by the gyroangle definition in Def. 4.2.   □

The following theorem captures obvious analogies, demonstrating that opposite gyroangles of a gyroparallelogram are congruent. The abstract gyroparallelogram is studied in Sec. 3.12 and its realization in an Einstein gyrovector plane and in a Möbius gyrovector plane are shown graphically, for instance, in Fig. 3.10, p. 98, and Fig. 3.11, p. 99.

Theorem 4.4. *Opposite sides of a gyroparallelogram are congruent, and opposite gyroangles of a gyroparallelogram are congruent.*

Proof. Let $ABDC$ be a gyroparallelogram, Fig. 3.9, p. 97. Its opposite sides BA and CD and, similarly, its opposite sides BD and CA, are related by (3.142), p. 94, so that we have

$$CD = \ominus C \oplus D = \ominus \text{gyr}[C, \ominus B](\ominus B \oplus A) = \ominus \text{gyr}[C, \ominus B]BA$$
$$CA = \ominus C \oplus A = \ominus \text{gyr}[C, \ominus B](\ominus B \oplus D) = \ominus \text{gyr}[C, \ominus B]BD.$$

(4.7)

Gyrations preserve the norm. Hence, the equations in (4.7) imply

$$|AB| := \|\ominus A \oplus B\| = \|\ominus C \oplus D\| =: |CD|$$
$$|AC| := \|\ominus A \oplus C\| = \|\ominus B \oplus D\| =: |BD|.$$

(4.8)

It follows from (4.8) that opposite sides of the gyroparallelogram are congruent.

Furthermore, it follows from the gyroangle definition, (4.2), and from (4.7), that

$$
\begin{aligned}
\cos \angle ACD &= \frac{\ominus C \oplus A}{\|\ominus C \oplus A\|} \cdot \frac{\ominus C \oplus D}{\|\ominus C \oplus D\|} \\
&= \frac{\ominus \text{gyr}[C, \ominus B](\ominus B \oplus D)}{\|\ominus \text{gyr}[C, \ominus B](\ominus B \oplus D)\|} \cdot \frac{\ominus \text{gyr}[C, \ominus B](\ominus B \oplus A)}{\|\ominus \text{gyr}[C, \ominus B](\ominus B \oplus A)\|} \\
&= \frac{\ominus B \oplus D}{\|\ominus B \oplus D\|} \cdot \frac{\ominus B \oplus A}{\|\ominus B \oplus A\|} \\
&= \cos \angle ABD,
\end{aligned}
$$

(4.9)

since gyrations preserve the inner product and the norm. It follows from (4.9) that opposite gyroangles of the gyroparallelogram are congruent. $\qquad \square$

The definition of $\cos \alpha$ in (4.2) is uniquely dictated by analogies with vector spaces. Remarkably, it is the gyroangle invariance under left gyrotranslations in (4.6) that dictates the form of the gyroangle definition in (4.2), in which we consider the gyrodifference $\ominus A \oplus B$, rather than the gyrodifference $B \ominus A$, as the gyro-analog of the difference $B - A = -A + B$ in vector spaces. A hypothetical gyroangle definition with $\ominus A \oplus B$ being replaced by $B \ominus A$, etc., would not be invariant under left gyrotranslations; see Remark 2.12, p. 38.

It follows from Theorem 4.3, and shown in Fig. 4.1, that a gyroangle α with vertex $A'' = \mathbf{0}$ at the origin and its gyrocosine, $\cos \alpha$, in a Möbius gyrovector space $(\mathbb{R}^n_s, \oplus, \otimes)$ behave like a Euclidean angle α and its cosine, $\cos \alpha$. Moreover, a gyroangle with any vertex in a Möbius gyrovector space $(\mathbb{R}^n_s, \oplus, \otimes)$ formed by two intersecting gyrolines remains equal to the Euclidean angle between corresponding tangent lines. In this sense a Möbius gyrovector space is said to be *conformal* to its corresponding Euclidean vector space. In contrast, only the origin of an Einstein gyrovector space is conformal to its corresponding Euclidean vector space.

As indicated in Figs. 4.2 – 4.3, the concept of gyrovector gyrotranslation, studied in Sec. 3.15, allows a natural extension of Def. 4.2 to the gyroangle between two gyrovectors that need not be

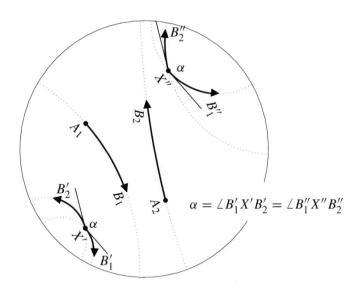

$$\alpha = \angle B_1' X' B_2' = \angle B_1'' X'' B_2''$$

Figure 4.2: Gyrovector gyrotranslations allow the visualization of the gyroangle formed by two rooted gyrovectors, $\ominus A_1 \oplus B_1$ and $\ominus A_2 \oplus B_2$, that have distinct tails, A_1 and A_2. The gyroangle is visually revealed by gyrovector translating the rooted gyrovectors into new rooted gyrovectors with a common tail X. The two cases of $X = X'$ and $X = X''$ in the Möbius gyrovector plane are shown here. The measure of the included gyroangle between two gyrovectors that share a common tail in a Möbius gyrovector space equals the Euclidean measure of the included angle between two tangent rays at the common tail. These tangent rays are, therefore, presented here.

rooted at a common point. Any two rooted gyrovectors that have no common tail can be gyrovector gyrotranslated to an arbitrarily given common tail. The resulting general gyroangle definition, illustrated in Figs. 4.2 – 4.3, follows.

Definition 4.5. **(Gyroangles Between Gyrovectors, II).** *Let $\ominus A_1 \oplus B_1$ and $\ominus A_2 \oplus B_2$ be two nonzero gyrovectors in a gyrovector space $(G, \oplus, \otimes)$. The gyrocosine of the gyroangle α, $0 \le \alpha \le \pi$, formed by the two gyrovectors, Figs. 4.2 – 4.3,*

$$\alpha = \angle(\ominus A_1 \oplus B_1)(\ominus A_2 \oplus B_2), \tag{4.10}$$

is given by the equation

$$\cos \alpha = \frac{\ominus A_1 \oplus B_1}{\|\ominus A_1 \oplus B_1\|} \cdot \frac{\ominus A_2 \oplus B_2}{\|\ominus A_2 \oplus B_2\|}. \tag{4.11}$$

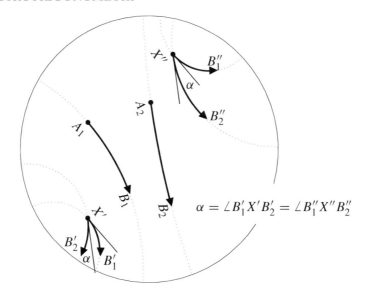

Figure 4.3:   Like Fig. 4.2, this figure illustrates the way gyroangles are formed by rooted gyrovectors that need not have a common tail. These are visualized here graphically in a Möbius gyrovector plane. The points of the Möbius gyrovector plane in this figure are the same points shown in the Möbius gyrovector plane of Fig. 4.2 with one exception: the points A_2 and B_2 have been interchanged.

4.2   THE GYROTRIANGLE

Any three non-gyrocollinear points A, B, C of a gyrovector space $(\mathbb{V}_s, \oplus, \otimes)$ form a gyrotriangle with vertices A, B, and C, denoted ABC, as shown in Fig. 3.15, p. 110, and in Fig. 4.4. The sides of a gyrotriangle ABC in the Einstein gyrovector plane $(\mathbb{R}_s^2, \oplus, \otimes)$, oriented counterclockwise, as shown in Fig. 4.4, form the gyrovectors

$$\mathbf{a} = \ominus B \oplus C, \qquad \mathbf{b} = \ominus C \oplus A, \qquad \mathbf{c} = \ominus A \oplus B. \qquad (4.12)$$

With a different orientation the sides of gyrotriangle ABC in Fig. 4.4 form other gyrovectors. These are

$$\mathbf{a}' = \ominus C \oplus B, \qquad \mathbf{b}' = \ominus A \oplus C, \qquad \mathbf{c}' = \ominus B \oplus A. \qquad (4.13)$$

Unlike vectors in Euclidean geometry, the oppositely oriented gyrovectors $\mathbf{a}$ and $\mathbf{a}'$ are not reciprocal to each other. Owing to the gyrocommutative law, the gyroautomorphic inverse property, Def. 2.1, p. 33, Theorem 1.13(12), p. 11, and the left loop property in Axiom (G5) of gyrogroups, the gyrovectors $\mathbf{a}$ and $\mathbf{a}'$ in (4.12) and (4.13) are related by identities that result from the following

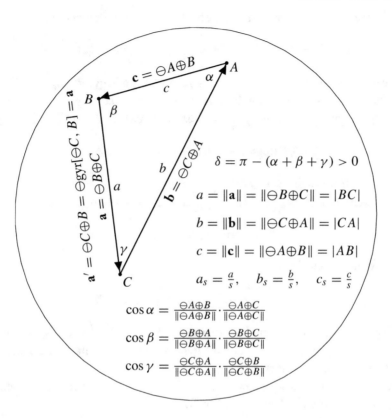

Figure 4.4: The gyrotriangle, and its standard notation, in an Einstein gyrovector space. The notation that we use with a gyrotriangle ABC, its gyrovector sides, and its gyroangles in an Einstein gyrovector space $(\mathbb{V}_s, \oplus, \otimes)$ is shown here for the Einstein gyrovector plane $(\mathbb{R}_s^2, \oplus, \otimes)$. The gyrotriangle defect δ is studied in Sec. 4.11. A gyrotriangle is *isosceles* if it has two sides congruent, and a gyrotriangle is *equilateral* if it has three sides congruent. We frequently use the same symbol, γ, for both a gyrotriangle gyroangle and the Lorentz factor. However, this should raise no confusion since the Lorentz factor always appears with a subindex.

chain of equations,

$$\begin{aligned}
\mathbf{a}' &= \ominus C \oplus B \\
&= \mathrm{gyr}[\ominus C, B](B \ominus C) \\
&= \ominus\mathrm{gyr}[\ominus C, B](\ominus B \oplus C) \\
&= \ominus\mathrm{gyr}[\ominus C, B]\mathbf{a} \\
&= \ominus\mathrm{gyr}[\ominus C \oplus B, B]\mathbf{a} \\
&= \ominus\mathrm{gyr}[\mathbf{a}', B]\mathbf{a}\,.
\end{aligned} \tag{4.14}$$

Hence, **a** and **a**′ have equal norms and, similarly, for **b** and **b**′, and for **c** and **c**′,

$$\|\mathbf{a}'\| = \|\mathbf{a}\|, \qquad \|\mathbf{b}'\| = \|\mathbf{b}\|, \qquad \|\mathbf{c}'\| = \|\mathbf{c}\|. \tag{4.15}$$

The three gyrotriangle side gyrovectors in (4.12)–(4.13) of a gyrotriangle ABC in a gyrovector space, like the ones shown in Fig. 4.4 for an Einstein gyrovector plane, with an appropriate orientation, satisfy the gyrogroup identity in Theorem 1.18, p. 16,

$$(\ominus A \oplus B) \oplus \mathrm{gyr}[\ominus A, B](\ominus B \oplus C) = \ominus A \oplus C. \tag{4.16}$$

We call (4.16) the *gyrotriangle (addition) law.*
Using the notation in (4.12)–(4.13), the gyrotriangle law (4.16) takes the form

$$\mathbf{c} \oplus \mathrm{gyr}[\ominus A, B]\mathbf{a} = \mathbf{b}', \tag{4.17}$$

or, equivalently, by the left loop property,

$$\mathbf{c} \oplus \mathrm{gyr}[\ominus A \oplus B, B]\mathbf{a} = \mathbf{c} \oplus \mathrm{gyr}[\mathbf{c}, B]\mathbf{a} = \mathbf{b}', \tag{4.18}$$

illustrated in Fig. 4.4. The gyrotriangle law, as presented in (4.17)–(4.18), is a gyrogroup, rather than a gyrovector, identity since it involves both gyrovectors and points. In Sec. 4.3, however, we will recast it into a gyrovector identity.
The three gyrotriangle side gyrovectors in (4.12)–(4.13), with appropriate orientations, also satisfy the gyrogroup identity in Theorem 2.11, p. 37,

$$\ominus(\ominus C \oplus B) \oplus (\ominus C \oplus A) = \mathrm{gyr}[\ominus C, B](\ominus B \oplus A), \tag{4.19}$$

or, equivalently, by the left loop property,

$$\ominus(\ominus C \oplus B) \oplus (\ominus C \oplus A) = \mathrm{gyr}[\ominus C \oplus B, B](\ominus B \oplus A), \tag{4.20}$$

$A, B, C \in \mathbb{V}_s$. In the notation (4.12)–(4.13), the latter, (4.20), can be written as

$$\mathbf{a}' \oplus \mathbf{b} = \mathrm{gyr}[\mathbf{a}', B]\mathbf{c}'. \tag{4.21}$$

The Euclidean analogs of the gyrogroup identities (4.16) and (4.19) are obvious. These are the vector space identities

$$(-A + B) + (-B + C) = -A + C \tag{4.22}$$

and

$$-(-C + B) + (-C + A) = -B + A, \tag{4.23}$$

$A, B, C \in \mathbb{V}$,

By Axiom $(V1)$ of gyrovector spaces, gyrations preserve the inner product and, hence, the norm. Hence, Identity (4.19) implies

$$\|\ominus(\ominus C \oplus B) \oplus (\ominus C \oplus A)\| = \|\ominus B \oplus A\| , \qquad (4.24)$$

$A, B, C \in \mathbb{V}_s$, thus obtaining a gyrovector space identity which has a form identical to its vector space counterpart,

$$\| - (-C + B) + (-C + A)\| = \| - B + A\| , \qquad (4.25)$$

$A, B, C \in \mathbb{V}$,

Identity (4.25) implies that Euclidean distance is invariant under Euclidean translations. In full analogy, Identity (4.24) implies that Einsteinian gyrodistance, Def. 3.10, p. 61, is invariant under left gyrotranslations.

We should note that in the Euclidean limit of large s, $s \to \infty$, the gyrogeometric identities (4.16), (4.19), and (4.24) reduce, respectively, to their Euclidean counterparts (4.22), (4.23), and (4.25).

4.3   THE GYROTRIANGLE ADDITION LAW

An application of the gyrotriangle law for gyrotriangle ABC in an Einstein gyrovector plane $(\mathbb{R}_s^2, \oplus, \otimes)$ given by (4.16)–(4.18) is illustrated in Fig. 4.4. The gyrotriangle law is presented in (4.16) as a gyrogroup identity, and in (4.17)–(4.18) as equivalent gyrogroup identities. In this section we will recast it into a gyrovector identity, that is, into an identity that involves only gyrovectors, as opposed to the gyrogroup identities in (4.16)–(4.18), which involve both gyrovectors and points.

Let us apply the gyrotriangle law (4.16) to gyrotriangle ABD in an Einstein gyrovector plane $(\mathbb{R}_s^2, \oplus, \otimes)$, illustrated in Fig. 4.5, obtaining in the notation of Fig. 4.5,

$$\begin{aligned}
\mathbf{w} = \ominus A \oplus D &= (\ominus A \oplus B) \oplus \mathrm{gyr}[\ominus A, B](\ominus B \oplus D) \\
&= (\ominus A \oplus B) \oplus \mathrm{gyr}[\ominus A \oplus B, B](\ominus B \oplus D) \qquad (4.26) \\
&= \mathbf{u} \oplus \mathrm{gyr}[\mathbf{u}, B] \oplus \mathbf{v}' .
\end{aligned}$$

In order to recast the gyrogroup identity (4.26) into a gyrovector identity, let us augment gyrotriangle ABD in Fig. 4.5 into gyroparallelogram $ABDC$, also shown in Fig. 4.5, by incorporating the vertex C given by the gyroparallelogram condition

$$C = (A \boxplus D) \ominus B . \qquad (4.27)$$

It follows from the gyroparallelogram condition (4.27) and Def. 3.38, p. 91, of the gyroparallelogram that the resulting gyroquadrilateral $ABDC$, shown in Fig. 4.5, is indeed a gyroparallelogram.

By Theorem 3.42, p. 94, with the notation in Fig. 4.5, side $\ominus B \oplus D$ of gyroparallelogram $ABDC$ is related to its opposite side $\ominus A \oplus C$ by the equation

$$\ominus B \oplus D = \mathrm{gyr}[B, \ominus C] \mathrm{gyr}[C, \ominus A](\ominus A \oplus C) . \qquad (4.28)$$

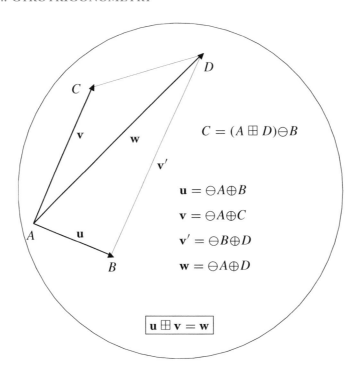

Figure 4.5: The gyrotriangle law, and the gyroparallelogram law of gyrovector addition, in an Einstein gyrovector space.

We now have the following chain of equations, which are numbered for subsequent derivation:

$$
\begin{aligned}
\ominus A \oplus D &\overset{(1)}{=\!=\!=} (\ominus A \oplus B) \oplus \mathrm{gyr}[\ominus A, B](\ominus B \oplus D) \\
&\overset{(2)}{=\!=\!=} (\ominus A \oplus B) \oplus \mathrm{gyr}[A, \ominus B]\mathrm{gyr}[B, \ominus C]\mathrm{gyr}[C, \ominus A](\ominus A \oplus C) \\
&\overset{(3)}{=\!=\!=} (\ominus A \oplus B) \oplus \mathrm{gyr}[\ominus A \oplus B, A \ominus C](\ominus A \oplus C) \\
&\overset{(4)}{=\!=\!=} (\ominus A \oplus B) \oplus \mathrm{gyr}[\ominus A \oplus B, \ominus(\ominus A \oplus C)](\ominus A \oplus C) \\
&\overset{(5)}{=\!=\!=} (\ominus A \oplus B) \boxplus (\ominus A \oplus C).
\end{aligned}
\tag{4.29}
$$

Derivation of the numbered equalities in (4.29) follows.

(1) Follows from the first equation in (4.26), which is the gyrotriangle law (4.16).

(2) Follows from (1) by substituting $\ominus B \oplus D$ from (4.28), and noting that, by the gyration even property (1.122), p. 27, $\mathrm{gyr}[\ominus A, B] = \mathrm{gyr}[A, \ominus B]$.

(3) Follows from (2) by Identity (2.31), p. 38.

(4) Follows from (3) by the gyroautomorphic inverse property, Def. 2.1, p. 33,

(5) Follows from (4) by the gyrogroup cooperation $\boxplus$, given by (1.20), p. 7.

Identities (4.26) and (4.29) provide the following two equivalent expressions for the gyrovector $\mathbf{w} = \ominus A \oplus D$ in Fig. 4.5,

$$\ominus A \oplus D = (\ominus A \oplus B) \oplus \text{gyr}[\ominus A \oplus B, B](\ominus B \oplus D)$$
$$\ominus A \oplus D = (\ominus A \oplus B) \boxplus (\ominus A \oplus C),$$

$$(4.30)$$

which, in the notation of Fig. 4.5, can be written as

$$\mathbf{w} = \mathbf{u} \oplus \text{gyr}[\mathbf{u}, B] \oplus \mathbf{v}'$$
$$\mathbf{w} = \mathbf{u} \boxplus \mathbf{v}.$$

$$(4.31)$$

The first identity in (4.31) gives the gyrotriangle law for gyrovector $\mathbf{w} = \ominus A \oplus D$ in gyrotriangle ABD of Fig. 4.5, and the second identity in (4.31) gives the gyroparallelogram law for gyrovector $\mathbf{w} = \ominus A \oplus D$ in gyroparallelogram $ABDC$ of Fig. 4.5. The gyrotriangle law and the gyroparallelogram law in (4.31) are equivalent. However, while the former is a gyrogroup identity that involves both gyrovectors and a point, the latter is a gyrovector identity between gyrovectors.

Thus, the attempt to recast the gyrotriangle law into an identity between gyrovectors, rather than an identity between gyrovectors and a point, results in the gyroparallelogram law. The latter, in turn, proves useful in the relativistic interpretation of the cosmological effect known as *stellar aberration*, which will be presented in Sec. 4.23.

The success of the relativistic interpretation of stellar aberration by means of the gyroparallelogram law in Einstein gyrovector spaces demonstrates that cosmological uniform velocities within the frame of Einstein's special theory of relativity are added according to the gyroparallelogram law of gyrovector addition, rather than by Einstein velocity addition law. Unlike Einstein addition of relativistically admissible velocities, which is neither commutative nor associative, the gyroparallelogram law of relativistically admissible velocities is commutative. Furthermore, in some specified general interpretation of associativity, the gyroparallelogram law of relativistically admissible velocities is also associative. In order to study the resulting generalized associativity, one must extend the gyroparallelogram law into a corresponding gyroparallelepiped law, a task that was accomplished in [61, Secs. 10.10 – 10.11].

Furthermore, the validity of the relativistic stellar aberration formulas, studied in [61, Chap. 13] and in the literature on special relativity, is central to the success of Gravity Probe B (GP-B), the NASA-Stanford University gyroscope experiment [11]. This experiment to measure the relativistic precession of gyroscopes in space is calibrated by means of relativistic stellar aberrations with an accuracy sufficient to achieve the desired overall accuracy.

Hence, the expected success of GP-B amounts to the expectation that the common relativistic stellar aberration formulas in the literature would be confirmed experimentally. This, in turn, amounts to the expectation that it would be confirmed experimentally that uniform velocities in the absence of gravitation are added according to the Einsteinian gyroparallelogram law of gyrovector addition, shown in Fig. 4.5, rather than by the standard Einstein velocity addition law. Hence, readers of this book are encouraged to read in [61, Chap. 13] how the common relativistic stellar aberration formulas that are obtained in the literature on special relativity by means of the Lorentz transformation, can be obtained by the Einstein gyroparallelogram, or gyrotriangle, law of gyrovector addition and by gyrotrigonometry in Einstein gyrovector spaces. The advantage of the latter way of obtaining the relativistic interpretation of stellar aberration is that it is guided by analogies with classical results of the standard vector space approach to Euclidean geometry.

To support the experimental validity of our result that uniform velocities in special relativity are added according to the gyroparallelogram law of gyrovector addition rather than according to Einstein velocity addition law, it is appropriate to remark here that the validity of Einstein velocity addition law as a matter of experimental fact has been established by Fizeau's 1851 experiment [31] only for the special case of parallel velocities. However, in this special case of parallel velocities the associated gyrations are trivial, so that both (i) Einstein addition ⊕, (2.64), p. 44, in Einstein addition law, and (ii) Einstein coaddition ⊞, (2.52), p. 42, in the Einsteinian gyroparallelogram law in Fig. 4.5, are identical to each other. They become distinguishable only in the addition of nonparallel velocities, when nontrivial gyrations emerge. Hence, in the classical Fizeau's experiment, Einstein's law and the Einsteinian gyroparallelogram law of velocity addition for parallel velocities are indistinguishable.

Indeed, the Fizeau experiment was carried out by Hippolyte Fizeau in the 1851 to measure the relative speeds of light in moving water. Albert Einstein later pointed out the importance of the experiment for special relativity, however, he did not mention it in his 1905 paper [9] that founded the special theory of relativity.

According to Kantor [23], generalized experiments of the Fizeau type to test Einstein velocity addition law for velocities that need not be parallel have already been reported with results that disagree with Einstein velocity addition law. Thus, by admitting the gyroparallelogram law in full analogy with its classical counterpart, the gyrogeometry that underlies Einstein's special theory of relativity reveals, once again, the connection between the mathematical beauty and simplicity of a theory and its truth. "A physical theory must possess mathematical beauty." That was the epigraph that Paul Dirac chose in 1956 when asked to express his view of the essence of physics. Accordingly, our gyrovector space approach to Einstein's special theory of relativity significantly extends Einstein's unfinished symphony.

4.4   COGYROLINES, COGYROTRIANGLES, AND COGYROANGLES

The gyroline definition in Def. 3.13, p. 62, is based on the left cancellation law as follows. The gyroline $\mathbf{a}\oplus(\ominus\mathbf{a}\oplus\mathbf{b})\otimes t$, $\mathbf{a}, \mathbf{b} \in \mathbb{V}_s$, $t \in \mathbb{R}$, is the unique one that passes through the points $\mathbf{a}$ and $\mathbf{b}$ of a gyrovector space $(\mathbb{V}_s, \oplus, \otimes)$. It passes through the point $\mathbf{a}$ when the gyroline parameter is given by $t = 0$ and, owing to the left cancellation law (1.70), p. 19, it passes through the point $\mathbf{b}$ when the gyroline parameter is given by $t = 1$.

The cogyroline definition is, similarly, based on a right cancellation law in the following definition.

Definition 4.6.   (Cogyrolines, Cogyrosegments). *Let* $\mathbf{a}, \mathbf{b}$ *be any two distinct points in a gyrovector space* $(\mathbb{V}_s, \oplus, \otimes)$. *The cogyroline in* $\mathbb{V}_s$ *that passes through the points* $\mathbf{a}$ *and* $\mathbf{b}$ *is the set of all points*

$$L = (\mathbf{b} \boxminus \mathbf{a}) \otimes t \oplus \mathbf{a} \tag{4.32}$$

in $\mathbb{V}_s$ *with the parameter* $t \in \mathbb{R}$. *The gyrovector space expression in* (4.32) *is called the representation of the cogyroline* L *in terms of the two points* $\mathbf{a}$ *and* $\mathbf{b}$ *that it contains.*

A cogyroline segment (or, a cogyrosegment) $\mathbf{ab}$ *with endpoints* $\mathbf{a}$ *and* $\mathbf{b}$ *is the set of all points in* (4.32) *with* $0 \leq t \leq 1$. *The cogyrolength* $|\mathbf{ab}|$ *of the cogyrosegment* $\mathbf{ab}$ *is the cogyrodistance between* $\mathbf{a}$ *and* $\mathbf{b}$,

$$|\mathbf{ab}| = d_{\boxplus}(\mathbf{a}, \mathbf{b}) = \| \boxminus \mathbf{a} \boxplus \mathbf{b} \| . \tag{4.33}$$

Two cogyrosegments are congruent if they have the same cogyrolength.

The cogyroline definition in Def. 4.6 is based on a right cancellation law as follows. The cogyroline $(\mathbf{b} \boxminus \mathbf{a}) \otimes t \oplus \mathbf{a}, \mathbf{a}, \mathbf{b} \in \mathbb{V}_s, t \in \mathbb{R}$, is the unique one that passes through the points $\mathbf{a}$ and $\mathbf{b}$ of a gyrovector space $(\mathbb{V}_s, \oplus, \otimes)$. It passes through the point $\mathbf{a}$ when the cogyroline parameter is given by $t = 0$ and, owing to the right cancellation law (1.71), p. 19, it passes through the point $\mathbf{b}$ when the cogyroline parameter is given by $t = 1$.

Following the definition of cogyrolines one may define cogyrotriangles and cogyroangles in gyrovector spaces in an obvious way. Thus, for instance, the three cogyroangles, α, β, and γ, of cogyrotriangle ABC in a gyrovector space $(\mathbb{V}_s, \oplus, \otimes)$ are given by the equations

$$
\begin{aligned}
\cos \alpha &= \frac{\boxminus A \boxplus B}{\| \boxminus A \boxplus B \|} \cdot \frac{\boxminus A \boxplus C}{\| \boxminus A \boxplus C \|} \\
\cos \beta &= \frac{\boxminus B \boxplus A}{\| \boxminus B \boxplus A \|} \cdot \frac{\boxminus B \boxplus C}{\| \boxminus B \boxplus C \|} \\
\cos \gamma &= \frac{\boxminus C \boxplus A}{\| \boxminus C \boxplus A \|} \cdot \frac{\boxminus C \boxplus B}{\| \boxminus C \boxplus B \|}
\end{aligned}
\tag{4.34}
$$

in full analogy with gyrotriangle gyroangles in Fig. 4.4, p. 117. Here $A \boxminus B = A \boxplus (\ominus B)$, so that $\boxminus A = \ominus A = -A$. Surprisingly, the cogyrotriangle cogyroangles α, β and γ in (4.34) satisfy the *Cogyrotriangle π Theorem*, [61, p. 365],

$$\alpha + \beta + \gamma = \pi . \tag{4.35}$$

Lines in Euclidean geometry admit the notion of parallelism. Surprisingly, in a similar way cogyrolines in hyperbolic geometry admit the notion of parallelism as well. Furthermore, triangle

angle sum in Euclidean geometry equals π. In a similar way cogyrotriangle cogyroangle sum in hyperbolic geometry equals π as well.

It is thus interesting to realize that parallelism does not disappear in the transition from Euclidean to hyperbolic geometry. While it disappears from the study of gyrolines, gyrotriangles and gyroangles of classical hyperbolic geometry, it reappears in the study of cogyrolines, cogyrotriangles, and cogyroangles in our gyrovector space approach to hyperbolic geometry. The study of cogyrolines, cogyrotriangles, and cogyroangles in gyrovector spaces is beyond the scope of this book. It is, however, found in [55] and [61].

4.5   THE LAW OF GYROCOSINES

By the notation in (4.12)–(4.13) for the sides of gyrotriangle ABC in an Einstein gyrovector space, Fig. 4.4, p. 117, and by (4.15), the gyrovector identity (4.24) can be written as

$$\|\ominus\mathbf{a}'\oplus\mathbf{b}\| = \|\mathbf{c}'\| = \|\mathbf{c}\| , \tag{4.36}$$

so that

$$\gamma_{\ominus\mathbf{a}'\oplus\mathbf{b}} = \gamma_{\mathbf{c}} . \tag{4.37}$$

It follows from (4.37) and the gamma identity (2.75), p. 46, that

$$\gamma_{\mathbf{c}} = \gamma_{\ominus\mathbf{a}'}\gamma_{\mathbf{b}}(1 - \frac{\mathbf{a}'\cdot\mathbf{b}}{s^2}) . \tag{4.38}$$

Noting that $\ominus\mathbf{a}'$ and $\mathbf{a}$ have equal magnitudes, a, Identity (4.38) can be written as

$$\gamma_c = \gamma_a\gamma_b(1 - \frac{\mathbf{a}'\cdot\mathbf{b}}{s^2}) = \gamma_a\gamma_b(1 - \frac{ab\cos\gamma}{s^2}) , \tag{4.39}$$

where, by Def. 4.2, p. 111, $\cos\gamma$ is the *gyrocosine* of gyroangle γ in Fig. 4.4, p. 117, satisfying the following equations:

$$\cos\gamma = \frac{\ominus C\oplus A}{\|\ominus C\oplus A\|}\cdot\frac{\ominus C\oplus B}{\|\ominus C\oplus B\|}$$
$$= \frac{\mathbf{a}'\cdot\mathbf{b}}{\|\mathbf{a}'\|\|\mathbf{b}\|} = \frac{\mathbf{a}'\cdot\mathbf{b}}{ab} , \tag{4.40}$$

and inequalities

$$-1 \le \cos\gamma \le 1 . \tag{4.41}$$

When $C = O$ is the origin of the Einstein gyrovector space $(\mathbb{V}_s, \oplus, \otimes)$, (4.40) reduces to

$$\cos\gamma = \frac{\ominus O\oplus A}{\|\ominus O\oplus A\|}\cdot\frac{\ominus O\oplus B}{\|\ominus O\oplus B\|}$$
$$= \frac{A}{\|A\|}\cdot\frac{B}{\|B\|} , \tag{4.42}$$

since the points A and B are identified with the gyrovectors $\ominus O \oplus A$ and $\ominus O \oplus B$, respectively.

Hence, a gyroangle with vertex at the origin, O, of its Einstein gyrovector space coincides with its Euclidean counterpart, as shown in Fig. 4.1, p. 112, for gyroangle $\angle B'' A'' C'' = \angle B'' O C''$.

It follows from (4.40) and (4.42) that the *gyrocosine* function cos is identical with the familiar Euclidean cosine function cos of a gyroangle γ in an Einstein gyrovector space. It is given by the inner product of two unit gyrovectors that emanate from a common point and have included gyroangle γ. Accordingly, it is justified to use the same notation, cos, for both the trigonometric cosine function and the gyrotrigonometric gyrocosine function.

Identity (4.39) for any gyrotriangle ABC, Fig. 4.4, p. 117, in any Einstein gyrovector space $(\mathbb{V}_s, \oplus, \otimes)$, written as

$$\gamma_c = \gamma_a \gamma_b (1 - a_s b_s \cos \gamma), \tag{4.43}$$

$a_s = a/s$, $b_s = b/s$, is called the *law of gyrocosines*. The law of gyrocosines (4.43) is equivalent to the Einstein gamma identity (4.38), thus providing a striking link between Einstein's special theory of relativity and the gyrotrigonometry of the hyperbolic geometry of Bolyai and Lobachevsky.

Remarkably, in the Euclidean limit of large s, $s \to \infty$, the law of gyrocosines (4.43) reduces to the trivial identity $1 = 1$. Hence, (4.43) has no immediate Euclidean counterpart, thus presenting a disanalogy between hyperbolic and Euclidean geometry. As a result, each of Theorems 4.7, p. 126, and 4.11, p. 129, below has no Euclidean counterpart as well.

As in trigonometry, the law of gyrocosines (4.43) is useful for calculating one side (that is, the side-gyrolength), c, of a gyrotriangle ABC, Fig. 4.4, p. 117, when the gyroangle γ opposite to side c and the other two sides, a and b, are known.

4.6   THE *SSS* TO *AAA* CONVERSION LAW

Let ABC be a gyrotriangle in an Einstein gyrovector space $(\mathbb{V}_s, \otimes, \oplus)$ with its standard notation in Fig. 4.4. According to (4.43), the gyrotriangle ABC possesses the following three identities, each of which represents its law of gyrocosines,

$$\begin{aligned} \gamma_a &= \gamma_b \gamma_c (1 - b_s c_s \cos \alpha) \\ \gamma_b &= \gamma_a \gamma_c (1 - a_s c_s \cos \beta) \\ \gamma_c &= \gamma_a \gamma_b (1 - a_s b_s \cos \gamma). \end{aligned} \tag{4.44}$$

Like Euclidean triangles, the gyroangles of a gyrotriangle are uniquely determined by its sides. Solving the system (4.44) of three identities for the three unknowns $\cos \alpha$, $\cos \beta$, and $\cos \gamma$, and employing (2.46), p. 41, we obtain the following theorem.

Theorem 4.7.   (The Law of Gyrocosines; The SSS to AAA Conversion Law). *Let ABC be a gyrotriangle in an Einstein gyrovector space* $(\mathbb{V}_s, \oplus, \otimes)$. *Then, in the gyrotriangle notation in Fig. 4.4, p. 117:*

$$\cos\alpha = \frac{-\gamma_a + \gamma_b\gamma_c}{\gamma_b\gamma_c b_s c_s} = \frac{-\gamma_a + \gamma_b\gamma_c}{\sqrt{\gamma_b^2 - 1}\sqrt{\gamma_c^2 - 1}}$$

$$\cos\beta = \frac{-\gamma_b + \gamma_a\gamma_c}{\gamma_a\gamma_c a_s c_s} = \frac{-\gamma_b + \gamma_a\gamma_c}{\sqrt{\gamma_a^2 - 1}\sqrt{\gamma_c^2 - 1}} \qquad (4.45)$$

$$\cos\gamma = \frac{-\gamma_c + \gamma_a\gamma_b}{\gamma_a\gamma_b a_s b_s} = \frac{-\gamma_c + \gamma_a\gamma_b}{\sqrt{\gamma_a^2 - 1}\sqrt{\gamma_b^2 - 1}}.$$

The identities in (4.45) form the SSS (Side-Side-Side) to AAA (gyroAngle-gyroAngle-gyroAngle) conversion law in Einstein gyrovector spaces. This law is useful for calculating the gyroangles of a gyrotriangle in an Einstein gyrovector space when its sides (that is, its side-gyrolengths) are known.

In full analogy with the trigonometry of triangles, the *gyrosine* of a gyrotriangle gyroangle α is nonnegative, given by the equation

$$\sin\alpha = \sqrt{1 - \cos^2\alpha}. \qquad (4.46)$$

Hence, it follows from Theorem 4.7 that the gyrosine of the gyrotriangle gyroangles in that Theorem are given by

$$\sin\alpha = \frac{\sqrt{1 + 2\gamma_a\gamma_b\gamma_c - \gamma_a^2 - \gamma_b^2 - \gamma_c^2}}{\sqrt{\gamma_b^2 - 1}\sqrt{\gamma_c^2 - 1}}$$

$$\sin\beta = \frac{\sqrt{1 + 2\gamma_a\gamma_b\gamma_c - \gamma_a^2 - \gamma_b^2 - \gamma_c^2}}{\sqrt{\gamma_a^2 - 1}\sqrt{\gamma_c^2 - 1}} \qquad (4.47)$$

$$\sin\gamma = \frac{\sqrt{1 + 2\gamma_a\gamma_b\gamma_c - \gamma_a^2 - \gamma_b^2 - \gamma_c^2}}{\sqrt{\gamma_a^2 - 1}\sqrt{\gamma_b^2 - 1}}.$$

Any gyrotriangle gyroangle α satisfies the inequality $0 < \alpha < \pi$, so that $\sin\alpha > 0$. Hence, following (4.47) we have the inequality

$$0 < 1 + 2\gamma_a\gamma_b\gamma_c - \gamma_a^2 - \gamma_b^2 - \gamma_c^2 \qquad (4.48)$$

for any gyrotriangle in an Einstein gyrovector space, in the notation of Theorem 4.7 and Fig. 4.4. Identities (4.47) immediately give rise to the identities

$$\frac{\sin\alpha}{\sqrt{\gamma_a^2 - 1}} = \frac{\sin\beta}{\sqrt{\gamma_b^2 - 1}} = \frac{\sin\gamma}{\sqrt{\gamma_c^2 - 1}}, \tag{4.49}$$

which form the law of gyrosines that we will study in Theorem 4.16, p. 133.

Example 4.8.   As an application of the gyrotriangle identities (4.45) and (4.47) we note that these identities imply the elegant identity

$$\frac{\sin\beta + \sin\gamma}{\sin(\beta + \gamma)} = \sqrt{\frac{\gamma_a + 1}{\gamma_a - 1}} \frac{\sqrt{\gamma_b^2 - 1} + \sqrt{\gamma_c^2 - 1}}{\gamma_b + \gamma_c}$$

$$= \frac{1}{s}\sqrt{\frac{\gamma_a + 1}{\gamma_a - 1}} \frac{\gamma_b\, b + \gamma_c\, c}{\gamma_b + \gamma_c} \tag{4.50}$$

for any gyrotriangle ABC in an Einstein gyrovector space.

In the gyrotriangle identity (4.50) we use the notation in Fig. 4.4, p. 117, and the trigonometric/gyrotrigonometric functional identity

$$\sin(\beta + \gamma) = \sin\beta\cos\gamma + \cos\beta\sin\gamma \tag{4.51}$$

for gyroangles β and γ.

Identity (4.50) will prove useful in the proof of the Breusch's Lemma 4.39, p. 157, in hyperbolic geometry.

4.7   INEQUALITIES FOR GYROTRIANGLES

Elegant inequalities for gyrotriangles in Einstein gyrovector spaces result immediately from (4.47), as we see in the following theorem.

Theorem 4.9.   *Let ABC be a gyrotriangle in an Einstein gyrovector space, with the notation in Fig. 4.4, p. 117, for its side gyrolengths a, b, c. Then*

$$\gamma_a^2 + \gamma_b^2 + \gamma_c^2 - 1 < 2\gamma_a\gamma_b\gamma_c \le \gamma_a^2 + \gamma_b^2\gamma_c^2$$

$$\gamma_a^2 + \gamma_b^2 + \gamma_c^2 - 1 < 2\gamma_a\gamma_b\gamma_c \le \gamma_b^2 + \gamma_a^2\gamma_c^2 \tag{4.52}$$

$$\gamma_a^2 + \gamma_b^2 + \gamma_c^2 - 1 < 2\gamma_a\gamma_b\gamma_c \le \gamma_c^2 + \gamma_a^2\gamma_b^2 .$$

Proof. The left inequality in each of the three chains of inequalities in (4.52) follows immediately from (4.48).

The inequality $\sin^2 \alpha \leq 1$ implies, by means of (4.47), the inequality

$$\frac{1 + 2\gamma_a \gamma_b \gamma_c - \gamma_a^2 - \gamma_b^2 - \gamma_c^2}{(\gamma_b^2 - 1)(\gamma_c^2 - 1)} \leq 1 \,, \tag{4.53}$$

which, in turn, implies

$$2\gamma_a \gamma_b \gamma_c \leq \gamma_a^2 + \gamma_b^2 \gamma_c^2 \,, \tag{4.54}$$

thus verifying the first chain of inequalities in (4.52). The proof of the other two chains of inequalities in (4.52) is similar. □

Equality is attained in Inequality (4.54) when, and only when, $\alpha = \pi/2$. In this case, gyrotriangle ABC is a right gyroangled gyrotriangle, satisfying the Einstein-Pythagoras identity $\gamma_a = \gamma_b \gamma_c$, (4.88), p. 136, that will be studied in Sec. 4.12.

Gyrotriangle gyroangles vary over the open interval $(0, \pi)$. Accordingly, we present the following definition about gyrotriangles and their gyroangles.

Definition 4.10.   (Acute, Right, Obtuse Gyroangles, and Gyrotriangles). *An acute gyroangle θ is a gyroangle measuring between 0 and $\pi/2$ radians, $0 < \theta < \pi/2$.*
A right gyroangle θ is a gyroangle measuring $\pi/2$ radians, $\theta = \pi/2$.
An obtuse gyroangle θ is a gyroangle measuring between $\pi/2$ and π radians, $\pi/2 < \theta < \pi$.
A gyrotriangle in which all three gyroangles are acute, $0 < \alpha, \beta, \gamma < \pi/2$, is acute.
A gyrotriangle in which one gyroangle is a right gyroangle, $\theta = \pi/2$, is right.
A gyrotriangle which has an obtuse gyroangle, $\pi/2 < \theta < \pi$, is obtuse.

It follows from (4.45), and from the result that the trigonometric cosine function, $\cos \alpha$, and the gyrotrigonometric gyrocosine function, ambiguously also denoted $\cos \alpha$, have the same behavior, that we have important equalities and inequalities for gyrotriangles in Einstein gyrovector spaces. These are:

$$\begin{cases} \gamma_b \, \gamma_c \; > \gamma_a & \text{if and only if } \alpha < \frac{\pi}{2} \text{ (Acute)}; \\ \gamma_b \, \gamma_c \; = \gamma_a & \text{if and only if } \alpha = \frac{\pi}{2} \text{ (Right)}; \\ \gamma_b \, \gamma_c \; < \gamma_a & \text{if and only if } \alpha > \frac{\pi}{2} \text{ (Obtuse)}, \end{cases} \tag{4.55}$$

and, similarly,

$$\begin{cases} \gamma_a \, \gamma_c \; > \gamma_b & \text{if and only if } \beta < \frac{\pi}{2} \text{ (Acute)}; \\ \gamma_a \, \gamma_c \; = \gamma_b & \text{if and only if } \beta = \frac{\pi}{2} \text{ (Right)}; \\ \gamma_a \, \gamma_c \; < \gamma_b & \text{if and only if } \beta > \frac{\pi}{2} \text{ (Obtuse)}, \end{cases} \tag{4.56}$$

and

$$\begin{cases} \gamma_a\,\gamma_b \;>\; \gamma_c & \text{if and only if } \gamma < \frac{\pi}{2} \text{ (Acute);} \\ \gamma_a\,\gamma_b \;=\; \gamma_c & \text{if and only if } \gamma = \frac{\pi}{2} \text{ (Right);} \\ \gamma_a\,\gamma_b \;<\; \gamma_c & \text{if and only if } \gamma > \frac{\pi}{2} \text{ (Obtuse),} \end{cases} \tag{4.57}$$

where **a**, **b**, **c** are the sides of any given gyrotriangle in an Einstein gyrovector space and α, β, γ are their respective opposing gyroangles, as shown in Fig. 4.4, p. 117. The equalities in (4.55)–(4.57) correspond to right gyroangles, giving rise to the Einstein-Pythagoras identity (4.88), p. 136, that will be studied in Sec. 4.12.

4.8   THE *AAA* TO *SSS* CONVERSION LAW

Unlike Euclidean triangles, the side gyrolengths of a gyrotriangle are uniquely determined by its gyroangles, as the following theorem demonstrates.

Theorem 4.11.   (The *AAA* to *SSS* Conversion Law). *Let ABC be a gyrotriangle in an Einstein gyrovector space* $(\mathbb{V}_s, \oplus, \otimes)$. *Then, in the gyrotriangle notation in Fig. 4.4, p. 117,*

$$\begin{aligned} \gamma_a &= \frac{\cos\alpha + \cos\beta\cos\gamma}{\sin\beta\sin\gamma} \\[4pt] \gamma_b &= \frac{\cos\beta + \cos\alpha\cos\gamma}{\sin\alpha\sin\gamma} \\[4pt] \gamma_c &= \frac{\cos\gamma + \cos\alpha\cos\beta}{\sin\alpha\sin\beta} \,, \end{aligned} \tag{4.58}$$

where, following (4.46), the gyrosine of the gyrotriangle gyroangle α, $\sin\alpha$, is the nonnegative value of $\sqrt{1 - \cos^2\alpha}$, etc.

Proof. Let ABC be a gyrotriangle in an Einstein gyrovector space $(\mathbb{V}_s, \otimes, \oplus)$ with its standard notation in Fig. 4.4. It follows straightforwardly from the *SSS* to *AAA* conversion law (4.45) that

$$\left(\frac{\cos\alpha + \cos\beta\cos\gamma}{\sin\beta\sin\gamma}\right)^2 = \frac{(\cos\alpha + \cos\beta\cos\gamma)^2}{(1 - \cos^2\beta)(1 - \cos^2\gamma)} = \gamma_a^2, \tag{4.59}$$

implying the first identity in (4.58). The remaining two identities in (4.58) are obtained from (4.45) in a similar way.   □

The identities in (4.58) form the *AAA* to *SSS* conversion law. This law is useful for calculating the sides (that is, the side-gyrolengths) of a gyrotriangle in an Einstein gyrovector space when its gyroangles are known. Thus, for instance, γ_a is obtained from the first identity in (4.58), and a is obtained from γ_a by Identity (2.46), p. 41.

Solving the third identity in (4.58) for $\cos \gamma$ we have

$$\cos \gamma = -\cos \alpha \cos \beta + \gamma_c \sin \alpha \sin \beta$$
$$= -\cos(\alpha + \beta) + (\gamma_c - 1) \sin \alpha \sin \beta , \tag{4.60}$$

implying

$$\cos \gamma = \cos(\pi - \alpha - \beta) + (\gamma_c - 1) \sin \alpha \sin \beta . \tag{4.61}$$

In the Euclidean limit of large s, $s \to \infty$, γ_c reduces to 1, so that the gyrotrigonometric identity (4.61) reduces to the trigonometric identity

$$\cos \gamma = \cos(\pi - \alpha - \beta) \tag{4.62}$$

in Euclidean geometry. The latter, in turn, is equivalent to the familiar result,

$$\alpha + \beta + \gamma = \pi , \tag{4.63}$$

of Euclidean geometry, according to which the triangle angle sum is π.

As an immediate application of the SSS to AAA conversion law (4.45) – (4.47) and the AAA to SSS conversion law (4.58) we present the following two theorems.

Theorem 4.12.   (The Isosceles Gyrotriangle Theorem). *A gyrotriangle is isosceles (that is, it has two sides congruent) if and only if the two gyroangles opposing its two congruent sides are congruent.*

Proof. Using the gyrotriangle notation in Fig. 4.4, p. 117, if gyrotriangle ABC has two sides, say a and b, congruent, $a = b$, then $\gamma_a = \gamma_b$ so that, by (4.45) – (4.47), $\cos \alpha = \cos \beta$ and $\sin \alpha = \sin \beta$ implying $\alpha = \beta$.

Conversely, if gyrotriangle ABC has two gyroangles, say α and β, congruent, then by (4.58), $\gamma_a = \gamma_b$, implying $a = b$.   □

Theorem 4.13.   (The Equilateral Gyrotriangle Theorem). *A gyrotriangle is equilateral (that is, it has all three sides congruent) if and only if the gyrotriangle is equigyroangular (that is, it has all three gyroangles congruent).*

Proof. Using the gyrotriangle notation in Fig. 4.4, p. 117, if gyrotriangle ABC has all three sides congruent, $a = b = c$, then $\gamma_a = \gamma_b = \gamma_c$ so that, by (4.45) – (4.47), $\cos \alpha = \cos \beta = \cos \gamma$ and $\sin \alpha = \sin \beta = \sin \gamma$, implying $\alpha = \beta = \gamma$.

Conversely, if gyrotriangle ABC has all three gyroangles congruent, then $\alpha = \beta = \gamma$ implying, by (4.58), $\gamma_a = \gamma_b = \gamma_c$. The latter implies $a = b = c$ by means of (2.46), p. 41.   □

Inequalities in (4.48) and (4.55)–(4.57) prove useful in establishing the following lemma.

Lemma 4.14. *One side of a gyrotriangle is greater than a second (that is, it has a greater gyrolength) if and only if the gyroangle opposite the first is greater (that is, it has a greater measure) than the gyroangle opposite the second.*

Proof. We prove the Lemma for Einstein gyrovector spaces. Let ABC be a gyrotriangle in a gyrovector space $(\mathbb{V}_s, \oplus, \otimes)$, shown in Fig. 4.4, p. 117, along with its standard notation that we use. We have to prove that $\beta > \gamma$ if and only if $b > c$.

Part (A): If $\beta > \gamma$ then $b > c$. The proof of Part (A) is divided into two cases.

Case I: Gyrotriangle ABC is obtuse, and $\beta > \pi/2$. Obviously, in this case we have $\beta > \gamma$. Furthermore, in this case it follows from (4.56) that

$$\gamma_b > \gamma_a \gamma_c \tag{4.64}$$

so that, by (4.48),

$$0 < 1 + 2\gamma_a\gamma_b\gamma_c - \gamma_a^2 - \gamma_b^2 - \gamma_c^2 \leq 1 + 2\gamma_b^2 - \gamma_a^2 - \gamma_b^2 - \gamma_c^2 = 1 - \gamma_a^2 + \gamma_b^2 - \gamma_c^2. \tag{4.65}$$

Hence,

$$\gamma_b^2 > \gamma_c^2 + (\gamma_a^2 - 1) > \gamma_c^2, \tag{4.66}$$

so that

$$\gamma_b > \gamma_c. \tag{4.67}$$

But, γ_x, $-s < x < s$, is a monotonically increasing function of x. Hence (4.67) implies $b > c$, as desired.

Case II: Gyrotriangle ABC is obtuse, and $\beta < \pi/2$, or gyrotriangle ABC is right or acute. We assume

$$\beta > \gamma. \tag{4.68}$$

In this case we have $0 < \beta, \gamma \leq \pi/2$, so that both $\sin\beta$ and $\sin\gamma$ are monotonically increasing functions. Hence, assumption (4.68) implies

$$\sin\beta > \sin\gamma. \tag{4.69}$$

But, it follows from (4.49) that

$$\frac{\sqrt{\gamma_b^2 - 1}}{\sqrt{\gamma_c^2 - 1}} = \frac{\sin\beta}{\sin\gamma}, \tag{4.70}$$

so that by (4.69)–(4.70), $\gamma_b^2 - 1 > \gamma_c^2 - 1$, implying $\gamma_b > \gamma_c$. Finally, as in (4.67), the latter implies $b > c$, as desired.

Part (B): If $b > c$ then $\beta > \gamma$. Let gyrotriangle ABC, Fig. 4.4, p. 117, be such that $b > c$. Either $\beta < \gamma$, $\beta = \gamma$, or $\beta > \gamma$.

If $\beta < \gamma$ then, by Part (A) above, $b < c$, thus contradicting the assumption $b > c$. If $\beta = \gamma$ then, by the Isosceles Gyrotriangle Theorem 4.12, $b = c$, thus contradicting the assumption $b > c$. Hence, $\beta > \gamma$.   □

Lemma 4.15.   *Let two sides of one gyrotriangle be congruent, respectively, to the two sides of a second gyrotriangle. Then, the measure of the third side of the first gyrotriangle is greater than the measure of the third side of the second gyrotriangle if and only if the measure of the included gyroangle of the two sides of the first gyrotriangle is greater than the measure of the included gyroangle of the two sides of the second gyrotriangle.*

Proof. We prove the Lemma for Einstein gyrovector spaces. Let ABC and $A'B'C'$ be two gyrotriangles in a gyrovector space $(\mathbb{V}_s, \oplus, \otimes)$, shown in Fig. 4.4, p. 117, along with the gyrotriangle standard notation that we use. The side gyrolengths of gyrotriangle ABC opposing vertices A, B, C are thus a, b, c and, similarly, the side gyrolengths of gyrotriangle $A'B'C'$ opposing vertices A', B', C' are a', b', c'. Furthermore, the gyroangles of gyrotriangle ABC opposing sides a, b, c are α, β, γ and, similarly, the gyroangles of gyrotriangle $A'B'C'$ opposing sides a', b', c' are α', β', γ'.

We assume that $a = a'$ and $c = c'$, which is equivalent to the assumption

$$\begin{aligned} \gamma_a &= \gamma_{a'} \\ \gamma_c &= \gamma_{c'} , \end{aligned}$$ (4.71)

and we have to prove that $\beta > \beta'$ if and only if $b > b'$.

By (4.45) and (4.71) we have

$$\begin{aligned} \cos \beta &= \frac{-\gamma_b + \gamma_a \gamma_c}{\sqrt{\gamma_a^2 - 1}\sqrt{\gamma_c^2 - 1}} \\ \cos \beta' &= \frac{-\gamma_{b'} + \gamma_a \gamma_c}{\sqrt{\gamma_a^2 - 1}\sqrt{\gamma_c^2 - 1}} , \end{aligned}$$ (4.72)

for $0 < \beta, \beta' < \pi$, so that

$$\cos \beta < \cos \beta' \Longleftrightarrow \gamma_b > \gamma_{b'} .$$ (4.73)

But, for $0 < \beta < \pi$, $\cos \beta$ is a decreasing function of β so that

$$\cos \beta < \cos \beta' \Longleftrightarrow \beta > \beta' .$$ (4.74)

Hence, by (4.73) and (4.74) we have

$$\beta > \beta' \Longleftrightarrow \gamma_b > \gamma_{b'} \Longleftrightarrow b > b' ,$$ (4.75)

as desired.   □

4.9   THE LAW OF GYROSINES

Theorem 4.16.   (The Law of Gyrosines). *Let ABC be a gyrotriangle in an Einstein gyrovector space* $(\mathbb{V}_s, \oplus, \otimes)$. *Then, in the gyrotriangle notation in Fig. 4.4, p. 117,*

$$\frac{\sin \alpha}{\gamma_a a} = \frac{\sin \beta}{\gamma_b b} = \frac{\sin \gamma}{\gamma_c c} = \frac{1}{s} \sqrt{\frac{1 + 2\gamma_a \gamma_b \gamma_c - \gamma_a^2 - \gamma_b^2 - \gamma_c^2}{(\gamma_a^2 - 1)(\gamma_b^2 - 1)(\gamma_c^2 - 1)}}. \qquad (4.76)$$

Proof. It follows from (4.45)–(4.46) and Identity (2.46), p. 41, that

$$\left(\frac{\sin \alpha}{\gamma_a \, a}\right)^2 = \frac{1}{s^2} \frac{1 - \cos^2 \alpha}{\gamma_a^2 - 1} = \frac{1}{s^2} \frac{1 + 2\gamma_a \gamma_b \gamma_c - \gamma_a^2 - \gamma_b^2 - \gamma_c^2}{(\gamma_a^2 - 1)(\gamma_b^2 - 1)(\gamma_c^2 - 1)}$$

$$\left(\frac{\sin \beta}{\gamma_b \, b}\right)^2 = \frac{1}{s^2} \frac{1 - \cos^2 \beta}{\gamma_b^2 - 1} = \frac{1}{s^2} \frac{1 + 2\gamma_a \gamma_b \gamma_c - \gamma_a^2 - \gamma_b^2 - \gamma_c^2}{(\gamma_a^2 - 1)(\gamma_b^2 - 1)(\gamma_c^2 - 1)} \qquad (4.77)$$

$$\left(\frac{\sin \gamma}{\gamma_c \, c}\right)^2 = \frac{1}{s^2} \frac{1 - \cos^2 \gamma}{\gamma_c^2 - 1} = \frac{1}{s^2} \frac{1 + 2\gamma_a \gamma_b \gamma_c - \gamma_a^2 - \gamma_b^2 - \gamma_c^2}{(\gamma_a^2 - 1)(\gamma_b^2 - 1)(\gamma_c^2 - 1)}.$$

The result (4.76) of the theorem follows immediately from (4.77). □

One should note that the extreme right-hand side of each equation in (4.77) is symmetric in a, b, c. Interestingly, it has no counterpart in Euclidean geometry.

The law of gyrosines (4.76), excluding the extreme right-hand side in (4.76), is fully analogous to the law of sines, to which it reduces in the Euclidean limit, $s \to \infty$, of large s, when gamma factors tend to 1.

4.10   THE *ASA* TO *SAS* CONVERSION LAW

Theorem 4.17.   (The ASA to SAS Conversion Law). *Let ABC be a gyrotriangle in an Einstein gyrovector space* $(\mathbb{V}_s, \oplus, \otimes)$. *Then, in the gyrotriangle notation in Fig. 4.4, p. 117,*

$$a_s = \frac{\gamma_c c_s \sin \alpha}{\cos \alpha \sin \beta + \gamma_c \sin \alpha \cos \beta}$$

$$b_s = \frac{\gamma_c c_s \sin \beta}{\cos \beta \sin \alpha + \gamma_c \sin \beta \cos \alpha} \qquad (4.78)$$

$$\cos \gamma = \cos(\pi - \alpha - \beta) + (\gamma_c - 1) \sin \alpha \sin \beta.$$

Proof. Noting (2.46), p. 41, and using Mathematica, it follows from the third identity in (4.58) and from (4.45) that

$$\left(\frac{\gamma_c c_s \sin\alpha}{\cos\alpha\sin\beta + \gamma_c\sin\alpha\cos\beta}\right)^2 = \left(\frac{\gamma_c c_s \sin\alpha\sin\beta}{\cos\alpha\sin^2\beta + \gamma_c\sin\alpha\sin\beta\cos\beta}\right)^2$$

$$= \frac{(\gamma_c^2 - 1)(1 - \cos^2\alpha)(1 - \cos^2\beta)}{(\cos\alpha(1 - \cos^2\beta) + (\cos\gamma + \cos\alpha\cos\beta)\cos\beta)^2} \tag{4.79}$$

$$= \frac{(\gamma_c^2 - 1)(1 - \cos^2\alpha)(1 - \cos^2\beta)}{(\cos\alpha + \cos\beta\cos\gamma)^2}$$

$$= a_s^2\,,$$

thus obtaining the first identity in (4.78). The second identity in (4.78), for b_s, is obtained from the first identity, for a_s, by interchanging α and β. Finally, the third identity in (4.78) has been verified in (4.61). □

Employing Identity (2.46), p. 41, for c_s^2, the first two identities in (4.78) imply

$$a_s b_s = \frac{(\gamma_c^2 - 1)\sin\alpha\sin\beta}{(\cos\alpha\sin\beta + \gamma_c\sin\alpha\cos\beta)(\cos\beta\sin\alpha + \gamma_c\sin\beta\cos\alpha)}\,. \tag{4.80}$$

This identity will prove useful in calculating the gyrotriangle defect in Sec. 4.11.

The system of identities (4.78) of Theorem 4.17 gives the ASA (gyroAngle-Side-gyroAngle) to SAS (Side-gyroAngle-Side) conversion law. It is useful for calculating two sides and their included gyroangle of a gyrotriangle in an Einstein gyrovector space when the remaining two gyroangles and the side included between them are known.

4.11   THE GYROTRIANGLE DEFECT

Some algebraic manipulations are too difficult to be performed by hand, but can easily be accomplished by computer algebra, that is, a computer software for symbolic manipulation, like Mathematica or Maple. It follows by straightforward substitution from (4.58) and (4.80), using computer algebra, that

$$\frac{\gamma_a\gamma_b a_s b_s \sin\gamma}{(1 + \gamma_a)(1 + \gamma_b) - \gamma_a\gamma_b a_s b_s \cos\gamma} = \cot\frac{\alpha + \beta + \gamma}{2}$$

$$= \tan\frac{\pi - (\alpha + \beta + \gamma)}{2} \tag{4.81}$$

$$= \tan\tfrac{\delta}{2}$$

where

$$\delta = \pi - (\alpha + \beta + \gamma) \tag{4.82}$$

is the gyrotriangular *defect* of the gyrotriangle ABC, presented in Fig. 4.4, p. 117.

Identity (4.81) is useful for calculating the defect of a gyrotriangle in an Einstein gyrovector space when two side gyrolengths and their included gyroangle of the gyrotriangle are known.

Let us now substitute $\cos \gamma$ from (4.45) and a_s^2 and b_s^2 from Identities like (2.46), p. 41, into $\tan^2(\delta/2)$ of (4.81), obtaining the identity

$$
\tan^2 \frac{\delta}{2} = \frac{\gamma_a^2 \gamma_b^2 a_s^2 b_s^2 (1 - \cos^2 \gamma)}{((1 + \gamma_a)(1 + \gamma_b) - \gamma_a \gamma_b a_s b_s \cos \gamma)^2}
$$
$$
= \frac{1 + 2\gamma_a \gamma_b \gamma_c - \gamma_a^2 - \gamma_b^2 - \gamma_c^2}{(1 + \gamma_a + \gamma_b + \gamma_c)^2} ,
\tag{4.83}
$$

which constitutes the following theorem.

Theorem 4.18.   (The Gyrotriangular Defect I). *Let ABC be a gyrotriangle in an Einstein gyrovector space $(\mathbb{V}_s, \oplus, \otimes)$ with the standard gyrotriangle notation in Fig. 4.4, p. 117. Then the gyrotriangular defect δ of ABC is given by the equation*

$$
\tan \frac{\delta}{2} = \frac{\sqrt{1 + 2\gamma_a \gamma_b \gamma_c - \gamma_a^2 - \gamma_b^2 - \gamma_c^2}}{1 + \gamma_a + \gamma_b + \gamma_c} .
\tag{4.84}
$$

In the Newtonian-Euclidean limit of large s, $s \to \infty$, gamma factors reduce to 1 so that (4.84) reduces to $\tan(\delta/2) = 0$ implying $\delta = 0$. Hence, the triangular defects of Euclidean triangles vanish, as expected.

Theorem 4.19.   (The Gyrotriangular Defect II). *Let ABC be a gyrotriangle in an Einstein gyrovector space $(\mathbb{V}_s, \oplus, \otimes)$, with the standard gyrotriangle notation in Fig. 4.4. Then the gyrotriangular defect δ of gyrotriangle ABC is given by the equation*

$$
\tan \frac{\delta}{2} = \frac{p \sin \gamma}{1 - p \cos \gamma} ,
\tag{4.85}
$$

where $\gamma, \delta, p > 0$, and

$$
p^2 = \frac{\gamma_a - 1}{\gamma_a + 1} \frac{\gamma_b - 1}{\gamma_b + 1} .
\tag{4.86}
$$

Proof. Employing (2.46), p. 41, the first equation in (4.83) can be written as (4.85) where $\gamma, \delta, p > 0$, and where p is given by (4.86).   □

The gyrotriangular defect formula (4.84) of Theorem 4.18 is useful for calculating the defect of a gyrotriangle in an Einstein gyrovector space when the three sides of the gyrotriangle are known.

The gyrotriangular defect formula (4.85) of Theorem 4.19 is useful for calculating the defect of a gyrotriangle in an Einstein gyrovector space when two side and their included gyroangle of the gyrotriangle are known.

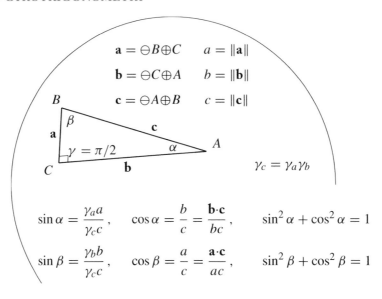

Figure 4.6:   Gyrotrigonometry in an Einstein gyrovector plane $(\mathbb{R}_s^2, \oplus, \otimes)$.

4.12   THE RIGHT GYROTRIANGLE

Let ABC be a right gyrotriangle in an Einstein gyrovector space $(\mathbb{V}_s, \oplus, \otimes)$ with the right gyroangle $\gamma = \pi/2$, as shown in Fig. 4.6 for $\mathbb{V}_s = \mathbb{R}_s^2$. It follows from (4.58) with $\gamma = \pi/2$ that the sides a, b and c of gyrotriangle ABC in Fig. 4.6 are related to the acute gyroangles α and β of the gyrotriangle by the equations

$$
\begin{aligned}
\gamma_a &= \frac{\cos \alpha}{\sin \beta} \\[2mm]
\gamma_b &= \frac{\cos \beta}{\sin \alpha} \\[2mm]
\gamma_c &= \frac{\cos \alpha \cos \beta}{\sin \alpha \sin \beta}.
\end{aligned}
\tag{4.87}
$$

The identities in (4.87) imply the Einstein-Pythagoras Identity

$$
\gamma_a \, \gamma_b = \gamma_c
\tag{4.88}
$$

for a right gyrotriangle ABC with hypotenuse c and legs a and b in an Einstein gyrovector space, Fig. 4.6. It follows from (4.88) that $\gamma_a^2 \gamma_b^2 = \gamma_c^2 = (1 - c_s^2)^{-1}$, implying that the gyrolength of the hypotenuse is given by the equation

$$
c_s = \frac{\sqrt{\gamma_a^2 \gamma_b^2 - 1}}{\gamma_a \gamma_b}.
\tag{4.89}
$$

In terms of *rapidities*, (4.88) takes the standard form,

$$\cosh\phi_a \cosh\phi_b = \cosh\phi_c \,, \tag{4.90}$$

of the hyperbolic Pythagorean theorem [18, p. 334][46, p. 151], where $\cosh\phi_a := \gamma_a$, etc.

In the special case of a right gyrotriangle in an Einstein gyrovector space, as in Fig. 4.6 where $\gamma = \pi/2$, it follows from Einstein-Pythagoras Identity (4.88) that the gyrotriangle defect identity (4.83) specializes to the right gyroangled gyrotriangle defect identity

$$\tan^2\frac{\delta}{2} = \frac{\gamma_a - 1}{\gamma_a + 1}\frac{\gamma_b - 1}{\gamma_b + 1} \,. \tag{4.91}$$

Identity (4.91) also follows from (4.85) with $\gamma = \pi/2$.

4.13 GYROTRIGONOMETRY

Let a, b, and c be the respective gyrolengths of the two legs **a**, **b** and the hypotenuse **c** of a right gyrotriangle ABC in an Einstein gyrovector space $(\mathbb{V}_s, \oplus, \otimes)$, Fig. 4.6. By (2.46), p. 41, and (4.87) we have

$$\left(\frac{a}{c}\right)^2 = \frac{(\gamma_a^2 - 1)/\gamma_a^2}{(\gamma_c^2 - 1)/\gamma_c^2} = \cos^2\beta$$

$$\left(\frac{b}{c}\right)^2 = \frac{(\gamma_b^2 - 1)/\gamma_b^2}{(\gamma_c^2 - 1)/\gamma_c^2} = \cos^2\alpha \,, \tag{4.92}$$

where γ_a , γ_b , and γ_c are related by (4.88).

Similarly, by (2.46), p. 41, and (4.87) we also have

$$\left(\frac{\gamma_a a}{\gamma_c c}\right)^2 = \frac{\gamma_a^2 - 1}{\gamma_c^2 - 1} = \sin^2\alpha$$

$$\left(\frac{\gamma_b b}{\gamma_c c}\right)^2 = \frac{\gamma_b^2 - 1}{\gamma_c^2 - 1} = \sin^2\beta \,. \tag{4.93}$$

Identities (4.92) and (4.93) imply

$$\left(\frac{a}{c}\right)^2 + \left(\frac{\gamma_b b}{\gamma_c c}\right)^2 = 1$$

$$\left(\frac{\gamma_a a}{\gamma_c c}\right)^2 + \left(\frac{b}{c}\right)^2 = 1 \,, \tag{4.94}$$

and, as shown in Fig. 4.6,

$$\cos\alpha = \frac{b}{c}$$
$$\cos\beta = \frac{a}{c}$$

(4.95)

and

$$\sin\alpha = \frac{\gamma_a a}{\gamma_c c}$$
$$\sin\beta = \frac{\gamma_b b}{\gamma_c c}\ .$$

(4.96)

Interestingly, we see from (4.95)–(4.96) that the gyrocosine function of an acute gyroangle of a right gyrotriangle in an Einstein gyrovector space has the same form as its Euclidean counterpart, the cosine function. In contrast, it is only modulo gamma factors that the gyrosine function has the same form as its Euclidean counterpart.

Identities (4.94) give rise to the following two distinct Einsteinian-Pythagorean identities,

$$a^2 + \left(\frac{\gamma_b}{\gamma_c}\right)^2 b^2 = c^2$$
$$\left(\frac{\gamma_a}{\gamma_c}\right)^2 a^2 + b^2 = c^2\ ,$$

(4.97)

for a right gyrotriangle with hypotenuse c and legs a and b in an Einstein gyrovector space. The two distinct Einsteinian-Pythagorean identities in (4.97) that each Einsteinian right gyrotriangle possesses converge in the Newtonian-Euclidean limit of large s, $s \to \infty$, to the single Pythagorean identity

$$a^2 + b^2 = c^2$$

(4.98)

that each Euclidean right-angled triangle possesses.

As an application of the gyrotrigonometry formed by (4.95)–(4.96) in Einstein gyrovector spaces, we verify the following theorem.

Theorem 4.20.   (The Base-Gyroaltitude Gyrotriangle Theorem). *Let ABC be a gyrotriangle with sides a, b, c in an Einstein gyrovector space $(\mathbb{V}_s, \oplus, \otimes)$, Fig. 4.4, p. 117, and let h_a, h_b, h_c be the three gyroaltitudes of ABC that correspond to gyroaltitudes drawn, respectively, from vertices A, B, C perpendicular to their opposite sides a, b, c or their extension (For instance, $h = h_c$ in Fig. 4.7). Then*

$$\gamma_a\, a\gamma_{h_a} h_a = \gamma_b\, b\gamma_{h_b} h_b = \gamma_c\, c\gamma_{h_c} h_c\ .$$

(4.99)

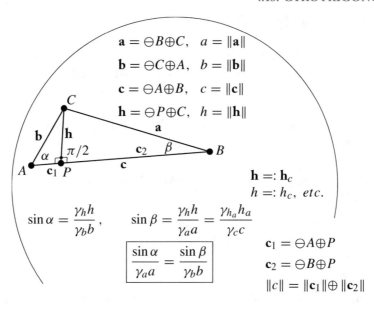

$$\mathbf{a} = \ominus B \oplus C, \quad a = \|\mathbf{a}\|$$
$$\mathbf{b} = \ominus C \oplus A, \quad b = \|\mathbf{b}\|$$
$$\mathbf{c} = \ominus A \oplus B, \quad c = \|\mathbf{c}\|$$
$$\mathbf{h} = \ominus P \oplus C, \quad h = \|\mathbf{h}\|$$

$$\mathbf{h} =: \mathbf{h}_c$$
$$h =: h_c, \ etc.$$

$$\sin \alpha = \frac{\gamma_h h}{\gamma_b b}, \qquad \sin \beta = \frac{\gamma_h h}{\gamma_a a} = \frac{\gamma_{h_a} h_a}{\gamma_c c}$$

$$\boxed{\frac{\sin \alpha}{\gamma_a a} = \frac{\sin \beta}{\gamma_b b}}$$

$$\mathbf{c}_1 = \ominus A \oplus P$$
$$\mathbf{c}_2 = \ominus B \oplus P$$
$$\|c\| = \|\mathbf{c}_1\| \oplus \|\mathbf{c}_2\|$$

Figure 4.7: The Law of Gyrosines. Drawing gyrotriangle gyroaltitudes in Einstein Gyrovector Spaces $(\mathbb{V}_s, \oplus, \otimes)$, and employing a basic gyrotrigonometric formula, shown in Fig. 4.6, we obtain the Law of Gyrosines, (4.76), in full analogy with the derivation of its Euclidean counterpart, the Law of Sines. The orthogonal projection of $\mathbf{a}$ on $\mathbf{c}' = \ominus B \oplus A$ is $\mathbf{c}_2$ and, similarly, the orthogonal projection of $\mathbf{b}' = \ominus A \oplus C$ on $\mathbf{c}$ is $\mathbf{c}_1$. Clearly, $\|\mathbf{b}'\| = \|\mathbf{b}\|$, and $\|\mathbf{c}'\| = \|\mathbf{c}\|$. It follows from the gyrotriangle equality, Theorem 3.28, p. 74, that $\|\mathbf{c}\| = \|\mathbf{c}_1\| \oplus \|\mathbf{c}_2\|$.

Proof. By (4.96) with the notation of Fig. 4.7 we have $\gamma_{h_a} h_a = \gamma_c \, c \sin \beta$. Hence, by (2.46), p. 41, (4.45), p. 126, and Identity (4.83), p. 135, for $\tan(\delta/2)$, we have the following chain of equations that culminates in a most unexpected, elegant result,

$$
\begin{aligned}
\gamma_a^2 a^2 \gamma_{h_a}^2 h_a^2 &= \gamma_a^2 a^2 \gamma_c^2 c^2 \sin^2 \beta \\
&= s^4 (\gamma_a^2 - 1)(\gamma_c^2 - 1)(1 - \cos^2 \beta) \\
&= s^4 (\gamma_a^2 - 1)(\gamma_c^2 - 1)(1 - \frac{(\gamma_a \gamma_c - \gamma_b)^2}{(\gamma_a^2 - 1)(\gamma_c^2 - 1)}) \qquad (4.100) \\
&= s^4 (1 + 2\gamma_a \gamma_b \gamma_c - \gamma_a^2 - \gamma_b^2 - \gamma_c^2) \\
&= s^4 (1 + \gamma_a + \gamma_b + \gamma_c)^2 \tan^2 \tfrac{\delta}{2},
\end{aligned}
$$

thus verifying the first identity in (4.101) below.

$$\begin{aligned}
\gamma_a a \gamma_{h_a} h_a &= s^2 (1 + \gamma_a + \gamma_b + \gamma_c) \tan(\delta/2) \\
\gamma_b b \gamma_{h_b} h_b &= s^2 (1 + \gamma_a + \gamma_b + \gamma_c) \tan(\delta/2) \\
\gamma_c c \gamma_{h_c} h_c &= s^2 (1 + \gamma_a + \gamma_b + \gamma_c) \tan(\delta/2) .
\end{aligned} \tag{4.101}$$

The remaining two identities in (4.101) follow from the first by interchanging a and b, and by interchanging a and c, respectively, noting that $\tan(\delta/2)$ is symmetric in a, b, c by Theorem 4.18, p. 135.

Finally, the *gyrotriangle base-gyroaltitude identity* (4.99) of the theorem follows from the three identities in (4.101).   □

Theorem 4.20 suggests the following definition.

Definition 4.21.   (The Gyrotriangle Constant). *Let a, b, c be the three sides of a gyrotriangle ABC with corresponding gyroaltitudes h_a, h_b, h_c in an Einstein gyrovector space $(\mathbb{V}_s, \oplus, \otimes)$, Figs. 4.4, p. 117, and 4.7.*

The number S_{ABC},

$$S_{ABC} = \gamma_a a \gamma_{h_a} h_a = \gamma_b b \gamma_{h_b} h_b = \gamma_c c \gamma_{h_c} h_c , \tag{4.102}$$

is called the gyrotriangle constant of gyrotriangle ABC.

It follows from Def. 4.21 and (4.100) that the gyrotriangle constant S_{ABC} of a gyrotriangle ABC with sides a, b, c and defect δ in any Einstein gyrovector space, Fig. 4.4, p. 117, satisfies each of the two identities

$$S_{ABC} = s^2 (1 + \gamma_a + \gamma_b + \gamma_c) \tan(\delta/2)$$

$$S_{ABC} = s^2 \sqrt{1 + 2\gamma_a \gamma_b \gamma_c - \gamma_a^2 - \gamma_b^2 - \gamma_c^2} . \tag{4.103}$$

Calling S_{ABC} in (4.103) a *gyrotriangle constant* of gyrotriangle ABC is justified since, as (4.103) demonstrates, S_{ABC} is invariant under any permutation of the sides a, b, c of the gyrotriangle ABC in Figs. 4.4 and 4.7.

We may note that following the elementary gyrotrigonometric identities shown in Figs. 4.6 and 4.7, the gyrotriangle constant (4.102) can be written in several forms. Thus, for instance,

$$S_{ABC} = \gamma_c c \gamma_{h_c} h_c = \gamma_c c \gamma_a a \sin \beta \tag{4.104}$$

for the gyrotriangle ABC in Fig. 4.7.

4.14   GYRODISTANCE BETWEEN A POINT AND A GYROLINE

Let A, B, $C \in \mathbb{R}_s^n$ be any three non-gyrocollinear points of an Einstein gyrovector space $(\mathbb{R}_s^n, \oplus, \otimes)$, $n \geq 3$, and let L_{AB} be the gyroline passing through the points A and B, Fig. 4.7. The gyrodistance between the point C and the gyroline L_{AB} is the gyrolength $h_c = \|\mathbf{h}_c\|$ of the perpendicular $\mathbf{h}_c$ from C to L_{AB}. With the notation of Fig. 4.7, and with permutations of a, b, c, it follows from (4.100) that

$$\gamma_a^2 a^2 \gamma_{h_a}^2 h_a^2 = \gamma_b^2 b^2 \gamma_{h_b}^2 h_b^2 = \gamma_c^2 c^2 \gamma_{h_c}^2 h_c^2$$
$$= s^4 (1 + 2\gamma_a \gamma_b \gamma_c - \gamma_a^2 - \gamma_b^2 - \gamma_c^2). \tag{4.105}$$

Hence, in particular, we have by the third equation in (4.105), and (2.46), p. 41,

$$\gamma_{h_c}^2 h_c^2 = s^2 \frac{1 + 2\gamma_a \gamma_b \gamma_c - \gamma_a^2 - \gamma_b^2 - \gamma_c^2}{\gamma_c^2 - 1}. \tag{4.106}$$

By an obvious identity and by (4.106) we have

$$\frac{1}{s^2} h_c^2 = 1 - \frac{1}{1 + \gamma_{h_c}^2 \frac{h_c^2}{s^2}}$$
$$= \frac{1 + 2\gamma_a \gamma_b \gamma_c - \gamma_a^2 - \gamma_b^2 - \gamma_c^2}{2\gamma_a \gamma_b \gamma_c - \gamma_a^2 - \gamma_b^2}, \tag{4.107}$$

so that, by (4.106) and (4.107),

$$\gamma_{h_c}^2 = \frac{2\gamma_a \gamma_b \gamma_c - \gamma_a^2 - \gamma_b^2}{\gamma_c^2 - 1}. \tag{4.108}$$

Formalizing the results in (4.105)–(4.108), we have the following theorem.

Theorem 4.22.   (Gyrodistance between a Point and a Gyroline). *Let A and B be any two points of an Einstein gyrovector space $(\mathbb{R}_s^n, \oplus, \otimes)$, $n \geq 3$, and let L_{AB} be the gyroline passing through these points. Furthermore, let C be any point of the space that does not lie on L_{AB}, as shown in Fig. 4.7. Then, in the notation of Fig. 4.7, the gyrodistance $h_c = \|\mathbf{h}_c\| = \|\ominus C \oplus P\|$ between the point C and the gyroline L_{AB} is given by the equation*

$$h_c^2 = s^2 \frac{1 + 2\gamma_a \gamma_b \gamma_c - \gamma_a^2 - \gamma_b^2 - \gamma_c^2}{2\gamma_a \gamma_b \gamma_c - \gamma_a^2 - \gamma_b^2} = s^2 (1 - \frac{\gamma_c^2 - 1}{2\gamma_a \gamma_b \gamma_c - \gamma_a^2 - \gamma_b^2}), \tag{4.109}$$

satisfying

$$\gamma_{h_c}^2 = \frac{2\gamma_a \gamma_b \gamma_c - \gamma_a^2 - \gamma_b^2}{\gamma_c^2 - 1}. \tag{4.110}$$

Furthermore, the product $\gamma_c c \gamma_{h_c} h_c$ is a symmetric function of a, b, c, given by the equation

$$\gamma_c^2 c^2 \gamma_{h_c}^2 h_c^2 = s^4 (1 + 2\gamma_a \gamma_b \gamma_c - \gamma_a^2 - \gamma_b^2 - \gamma_c^2). \tag{4.111}$$

As an immediate consequence of Theorem 4.22 we have the following Corollary.

Corollary 4.23.   Let A, B, C be any three non-gyrocollinear points of an Einstein gyrovector space, and let a, b, c be the gyrolengths of the sides of the gyrotriangle ABC,

$$\begin{aligned} a &= \|\ominus B \oplus C\| \\ b &= \|\ominus A \oplus C\| \\ c &= \|\ominus A \oplus B\| . \end{aligned} \tag{4.112}$$

Then, the gamma factors of a, b, c satisfy the following inequalities:

$$\begin{aligned} 2\gamma_a \gamma_b \gamma_c &> \gamma_a^2 + \gamma_b^2 \\ 1 + 2\gamma_a \gamma_b \gamma_c &> \gamma_a^2 + \gamma_b^2 + \gamma_c^2 . \end{aligned} \tag{4.113}$$

The second inequality in (4.113) reduces to its corresponding equality

$$1 + 2\gamma_a \gamma_b \gamma_c = \gamma_a^2 + \gamma_b^2 + \gamma_c^2 , \tag{4.114}$$

if and only if the points A, B, C are gyrocollinear.

Proof. If A, B, C are non-gyrocollinear then h_c is the gyrodistance between C and the line L_{AB} that passes through A and B, satisfying $h_c > 0$ and $\gamma_{h_c} > 1$. Hence, Inequalities (4.113) follow from (4.110) and from (4.111). Clearly, the second inequality in (4.113) reduces to equality (4.114) if and only if $h_c = 0$, that is, if and only if A, B, C are gyrocollinear. The second inequality in (4.113) was also verified in (4.48), p. 126. $\qquad\square$

The *orthogonal projection* of a point C onto a gyrosegment AB (or its extension, the gyroline L_{AB} that passes through the points A and B) is the foot P of the perpendicular CP from the point to the gyrosegment (or its extension), as shown in Fig. 4.7.

Accordingly, the orthogonal projection of side $\mathbf{b}' = \ominus A \oplus C$ on side $\mathbf{c} = \ominus A \oplus B$ of gyrotriangle ABC in Fig. 4.7 is

$$\mathbf{c}_1 = \ominus A \oplus P , \tag{4.115}$$

the gyrolength of which is

$$c_1 = \|\mathbf{c}_1\| = \|\ominus A \oplus P\| = b \cos \alpha , \tag{4.116}$$

as we see from Fig. 4.7 and from the relativistic gyrotrigonometry in Fig. 4.6, p. 136.

Hence, by (4.116), (2.46), p. 41, and (4.45), p. 126, we have

$$c_1^2 = b^2 \cos^2 \alpha = s^2 \frac{\gamma_b^2 - 1}{\gamma_b^2} \frac{(\gamma_b \gamma_c - \gamma_a)^2}{(\gamma_b^2 - 1)(\gamma_c^2 - 1)} = s^2 \frac{(\gamma_b \gamma_c - \gamma_a)^2}{\gamma_b^2 (\gamma_c^2 - 1)}, \tag{4.117}$$

so that

$$\gamma_{c_1}^2 = \gamma_b^2 \frac{\gamma_c^2 - 1}{2\gamma_a \gamma_b \gamma_c - \gamma_a^2 - \gamma_b^2} \tag{4.118}$$

Furthermore, following (4.117) and (4.55), p. 128, we have

$$c_1 = \begin{cases} s \dfrac{\gamma_b \, \gamma_c \, - \gamma_a}{\gamma_b \, \sqrt{\gamma_c^2 - 1}}, & \text{if } 0 < \alpha \leq \frac{\pi}{2}; \\[3ex] s \dfrac{\gamma_a \, - \gamma_b \, \gamma_c}{\gamma_b \, \sqrt{\gamma_c^2 - 1}}, & \text{if } \frac{\pi}{2} \leq \alpha < \pi. \end{cases} \tag{4.119}$$

It follows from (4.110) and (4.118) that

$$\gamma_{h_c}^2 \gamma_{c_1}^2 = \gamma_b^2, \tag{4.120}$$

in accordance with Einstein-Pythagoras Identity (4.88) for the right gyrotriangle APC in Fig. 4.7.

Similarly, the orthogonal projection of side $\mathbf{a} = \ominus B \oplus C$ on side $\mathbf{c}' = \ominus B \oplus A$ of gyrotriangle ABC in Fig. 4.7 is

$$\mathbf{c}_2 = \ominus B \oplus P, \tag{4.121}$$

the gyrolength of which is

$$c_2 = \|\mathbf{c}_2\| = \|\ominus B \oplus P\| = a \cos \beta, \tag{4.122}$$

as we see from Fig. 4.7 and from the basic equations of relativistic gyrotrigonometry in Fig. 4.6, p. 136.

Hence, by (4.122), (2.46), p. 41, and (4.45), we have

$$c_2^2 = a^2 \cos^2 \beta = s^2 \frac{\gamma_a^2 - 1}{\gamma_a^2} \frac{(\gamma_a \gamma_c - \gamma_b)^2}{(\gamma_a^2 - 1)(\gamma_c^2 - 1)} = s^2 \frac{(\gamma_a \gamma_c - \gamma_b)^2}{\gamma_a^2 (\gamma_c^2 - 1)}, \tag{4.123}$$

so that

$$\gamma_{c_2}^2 = \gamma_a^2 \frac{\gamma_c^2 - 1}{2\gamma_a \gamma_b \gamma_c - \gamma_a^2 - \gamma_b^2}. \tag{4.124}$$

Furthermore, following (4.123) and (4.56), p. 128, we have

$$c_2 = \begin{cases} s \dfrac{\gamma_a \, \gamma_c \, - \gamma_b}{\gamma_a \, \sqrt{\gamma_c^2 - 1}}, & \text{if } 0 < \beta \leq \frac{\pi}{2}; \\[3ex] s \dfrac{\gamma_b \, - \gamma_a \, \gamma_c}{\gamma_a \, \sqrt{\gamma_c^2 - 1}}, & \text{if } \frac{\pi}{2} \leq \beta < \pi. \end{cases} \tag{4.125}$$

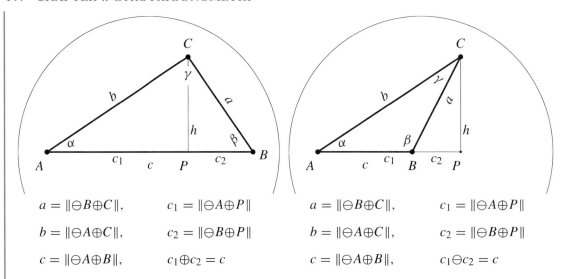

$$a = \|\ominus B \oplus C\|, \qquad c_1 = \|\ominus A \oplus P\|$$
$$b = \|\ominus A \oplus C\|, \qquad c_2 = \|\ominus B \oplus P\|$$
$$c = \|\ominus A \oplus B\|, \qquad c_1 \oplus c_2 = c$$

$$a = \|\ominus B \oplus C\|, \qquad c_1 = \|\ominus A \oplus P\|$$
$$b = \|\ominus A \oplus C\|, \qquad c_2 = \|\ominus B \oplus P\|$$
$$c = \|\ominus A \oplus B\|, \qquad c_1 \ominus c_2 = c$$

Figure 4.8: Orthogonal Projection I. Illustrating Example 4.24: Gyroangles α and β of a gyrotriangle ABC in an Einstein gyrovector space $(\mathbb{V}_s, \oplus, \otimes)$ are acute. As a result, $c = c_1 \oplus c_2$.

Figure 4.9: Orthogonal Projection II. Illustrating Example 4.25: Gyroangle β of a gyrotriangle ABC in an Einstein gyrovector space $(\mathbb{V}_s, \oplus, \otimes)$ is obtuse. As a result, $c = c_1 \ominus c_2$.

It follows from (4.110) and (4.124) that

$$\gamma_{h_c}^2 \gamma_{c_2}^2 = \gamma_a^2 , \tag{4.126}$$

in accordance with Einstein-Pythagoras Identity (4.88) for the right gyrotriangle BCP in Fig. 4.7.

Example 4.24. Let ABC be a gyrotriangle in an Einstein gyrovector space with acute gyroangles α and β, as shown in Fig. 4.8.

Following (4.117) and (4.55) for the acute gyroangle α in Fig. 4.8, we have

$$c_1 = s\frac{\gamma_b \, \gamma_c \, - \gamma_a}{\gamma_b \, \sqrt{\gamma_c^2 - 1}} \, . \tag{4.127}$$

Similarly, following (4.123), and (4.56) for the acute gyroangle β in Fig. 4.8, we have

$$c_2 = s\frac{\gamma_a \, \gamma_c \, - \gamma_b}{\gamma_a \, \sqrt{\gamma_c^2 - 1}} \, . \tag{4.128}$$

Hence, by (2.76), p. 46, by (4.127)–(4.128), and by (2.46), p. 41, we have

$$\frac{1}{s^2}(c_1 \oplus c_2)^2 = \frac{1}{s^2}\left(\frac{c_1 + c_2}{1 + \frac{c_1 c_2}{s^2}}\right)^2 = \frac{\gamma_c^2 - 1}{\gamma_c^2} = \frac{1}{s^2}c^2 , \tag{4.129}$$

implying

$$c_1 \oplus c_2 = c, \tag{4.130}$$

as expected in Fig. 4.8.

Example 4.25. Let ABC be a gyrotriangle in an Einstein gyrovector space with an acute gyroangle α and an obtuse gyroangle β, as shown in Fig. 4.9.

Following (4.117), and (4.55) for the acute gyroangle α in Fig. 4.9, we have

$$c_1 = s \frac{\gamma_b \; \gamma_c \; - \gamma_a}{\gamma_b \; \sqrt{\gamma_c^2 - 1}} \; . \tag{4.131}$$

Similarly, following (4.123), and (4.56) for the obtuse gyroangle β in Fig. 4.9, we have

$$c_2 = -s \frac{\gamma_a \; \gamma_c \; - \gamma_b}{\gamma_a \; \sqrt{\gamma_c^2 - 1}} \; . \tag{4.132}$$

Note that, unlike (4.128), the right-hand side of (4.132) is preceded by a negative sign.

Hence, by (2.76), p. 46, by (4.131)−(4.132), and by (2.46), p. 41, we have

$$\frac{1}{s^2}(c_1 \ominus c_2)^2 = \frac{1}{s^2} \left(\frac{c_1 - c_2}{1 - \frac{c_1 c_2}{s^2}} \right)^2 = \frac{\gamma_c^2 - 1}{\gamma_c^2} = \frac{1}{s^2} c^2 , \tag{4.133}$$

implying

$$c_1 \ominus c_2 = c , \tag{4.134}$$

as expected in Fig. 4.9.

Example 4.26. Let ABC be a gyrotriangle in an Einstein gyrovector space with acute gyroangles α and β, as shown in Fig. 4.8. For the product of the two orthogonal projections c_1 and c_2 on **c** we have, from (4.127)−(4.128),

$$c_1 c_2 = s^2 \frac{(\gamma_b \gamma_c - \gamma_a)(\gamma_a \gamma_c - \gamma_b)}{\gamma_a \gamma_b (\gamma_c^2 - 1)} , \tag{4.135}$$

and from (4.124) and (4.118),

$$\gamma_{c_1} \gamma_{c_2} = \gamma_a \gamma_b \frac{\gamma_c^2 - 1}{2 \gamma_a \gamma_b \gamma_c - \gamma_a^2 - \gamma_b^2} \; . \tag{4.136}$$

Hence, by Einstein gamma identity (2.74), p. 46, and by (4.135) and (4.136), we have

$$\gamma_{c_1 \oplus c_2} = \gamma_{c_1} \gamma_{c_2} (1 + \frac{c_1 c_2}{s^2}) = \gamma_c . \tag{4.137}$$

Indeed, identity (4.137) is expected from (4.130).

Example 4.27. Let ABC be a gyrotriangle in an Einstein gyrovector space with an acute gyroangle α and an obtuse gyroangle β, as shown in Fig. 4.9. For the product of the two orthogonal projections c_1 and c_2 on $\mathbf{c}$ we have, from (4.131)–(4.132),

$$c_1 c_2 = -s^2 \frac{(\gamma_b \gamma_c - \gamma_a)(\gamma_a \gamma_c - \gamma_b)}{\gamma_a \gamma_b (\gamma_c^2 - 1)}, \tag{4.138}$$

and from (4.124) and (4.118),

$$\gamma_{c_1} \gamma_{c_2} = \gamma_a \gamma_b \frac{\gamma_c^2 - 1}{2\gamma_a \gamma_b \gamma_c - \gamma_a^2 - \gamma_b^2}. \tag{4.139}$$

Hence, by Einstein gamma identity (2.75), p. 46, and by (4.138) and (4.139), we have

$$\gamma_{c_1 \ominus c_2} = \gamma_{c_1} \gamma_{c_2} (1 - \frac{c_1 c_2}{s^2}) = \gamma_c. \tag{4.140}$$

Indeed, identity (4.140) is expected from (4.134).

4.15   THE GYROTRIANGLE GYROALTITUDE

As an illustrative example of the application of gyrotrigonometry, we calculate in this section the gyroaltitude of a gyrotriangle in an Einstein gyrovector space, Fig. 4.10.

As shown in Fig. 4.10 for $\mathbb{V}_s = \mathbb{R}_s^2$, let ABC be a gyrotriangle in an Einstein gyrovector space $(\mathbb{V}_s, \oplus, \otimes)$, and let point $P \in \mathbb{V}_s$ be the orthogonal projection of vertex C onto its opposite side AB, so that gyrovector $\mathbf{h} = \ominus P \oplus C$ is the gyrotriangle gyroaltitude drawn from the gyrotriangle side AB. Then, by the gyrotrigonometric basic relations (4.95), illustrated in Fig. 4.6, the resulting right gyrotriangles APC and BPC in Fig. 4.10 give rise to the equations

$$\sin \alpha = \frac{\gamma_h h}{\gamma_b b}$$

$$\sin \beta = \frac{\gamma_h h}{\gamma_a a}, \tag{4.141}$$

in the notation of Fig. 4.10.

Hence,

$$\gamma_h h = \gamma_b b \sin \alpha = \gamma_a a \sin \beta, \tag{4.142}$$

thus recovering part of the law of gyrosines in Theorem 4.16, p. 133,

$$\frac{\sin \alpha}{\gamma_a a} = \frac{\sin \beta}{\gamma_b b}. \tag{4.143}$$

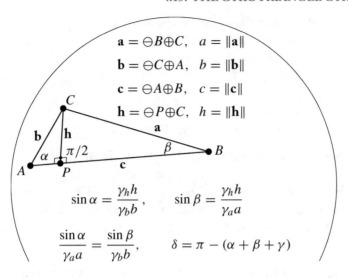

$$\mathbf{a} = \ominus B \oplus C, \quad a = \|\mathbf{a}\|$$
$$\mathbf{b} = \ominus C \oplus A, \quad b = \|\mathbf{b}\|$$
$$\mathbf{c} = \ominus A \oplus B, \quad c = \|\mathbf{c}\|$$
$$\mathbf{h} = \ominus P \oplus C, \quad h = \|\mathbf{h}\|$$

$$\sin \alpha = \frac{\gamma_h h}{\gamma_b b}, \qquad \sin \beta = \frac{\gamma_h h}{\gamma_a a}$$

$$\frac{\sin \alpha}{\gamma_a a} = \frac{\sin \beta}{\gamma_b b}, \qquad \delta = \pi - (\alpha + \beta + \gamma)$$

Figure 4.10: Gyrotriangle ABC and its gyroaltitude $\mathbf{h} = \ominus P \oplus C$ in a Möbius gyrovector plane $(\mathbb{R}_s^2, \oplus, \otimes)$ are shown, where point P is the orthogonal projection of vertex C onto its opposite side. The gyrolength $h = \|\mathbf{h}\|$ of the gyroaltitude $\mathbf{h}$ is calculated by means of gyrotrigonometry, giving rise to the Law of Gyrosines.

The product of the "relativistically corrected" gyroaltitude gyrolength $\gamma_h h$ of gyroaltitude $\mathbf{h}$ and the side gyrolength $\gamma_c c$ of side $\mathbf{c}$, the side over which gyroaltitude $\mathbf{h}$ is drawn as shown in Fig. 4.10, plays an important role in the gyrotriangle gyroarea that we will uncover in Sec. 4.16. The square of this product is given by the following chain of equations, which are numbered for subsequent explanation:

$$
\gamma_c^2 c^2 \gamma_h^2 h^2 \overset{(1)}{=\!=\!=} \gamma_b^2 b^2 \gamma_c^2 c^2 \sin^2 \alpha
$$
$$
\overset{(2)}{=\!=\!=} s^4 (\gamma_b^2 - 1)(\gamma_c^2 - 1)(1 - \cos^2 \alpha)
$$
$$
\overset{(3)}{=\!=\!=} s^4 (\gamma_b^2 - 1)(\gamma_c^2 - 1) \left(1 - \frac{(\gamma_b \gamma_c - \gamma_a)^2}{(\gamma_b^2 - 1)(\gamma_c^2 - 1)} \right) \tag{4.144}
$$
$$
\overset{(4)}{=\!=\!=} s^4 (1 + 2\gamma_a \gamma_b \gamma_c - \gamma_a^2 - \gamma_b^2 - \gamma_c^2)
$$
$$
\overset{(5)}{=\!=\!=} s^4 (1 + \gamma_a + \gamma_b + \gamma_c)^2 \tan^2 \tfrac{\delta}{2} .
$$

Derivation of the numbered equalities in (4.6) follows.

(1) Follows immediately from (4.142).

(2) Follows from (1) by (2.46), p. 41, and (4.46), p. 126.

(3) Follows from (2) by the first equation in (4.45), p. 126.

(4) Follows from (3) by straightforward algebra.

(5) Follows from (4) by (4.83), p. 135.

Hence, by (4.144),

$$\gamma_c c \gamma_h h = s^2 (1 + \gamma_a + \gamma_b + \gamma_c) \tan \tfrac{\delta}{2} \,, \tag{4.145}$$

where δ is the gyrotriangular defect of gyrotriangle ABC in Fig. 4.10.

Being a symmetric function of its gyrotriangle gyroangles, the gyrotriangular defect δ of a gyrotriangle ABC is a symmetric function of the gyrotriangle side gyrolengths a, b and c. Hence, the extreme right-hand side of (4.144) is a symmetric function of a, b, and c as well. This result implies that the left-hand side of (4.144) is a symmetric function of a, b, and c as well, thus obtaining the identities in Def. 4.21,

$$\gamma_a a \gamma_{h_a} h_a = \gamma_b b \gamma_{h_b} h_b = \gamma_c c \gamma_{h_c} h_c = s^2 (1 + \gamma_a + \gamma_b + \gamma_c) \tan \tfrac{\delta}{2} \,, \tag{4.146}$$

where $h_c = h$ is the gyroaltitude of gyrotriangle ABC drawn from side **c**, as shown in Fig. 4.10, and where h_a and h_b are the gyroaltitudes of gyrotriangle ABC drawn from sides **a** and **b**, respectively.

4.16   THE GYROTRIANGLE GYROAREA

Definition 4.28.   **(The Gyrotriangle Gyroarea).** Let ABC be a gyrotriangle in an Einstein gyrovector space $(\mathbb{V}_s, \oplus, \otimes)$. The gyroarea $|ABC|$ of gyrotriangle ABC is given by the equation

$$|ABC| = 2s^2 \tan \tfrac{\delta}{2} \,, \tag{4.147}$$

where δ is the defect of gyrotriangle ABC.

Theorem 4.29.   **(The Gyrotriangle Gyroarea).** *Let ABC be a gyrotriangle in an Einstein gyrovector space $(\mathbb{V}_s, \oplus, \otimes)$. Then, in the notation of Fig. 4.10, the gyroarea $|ABC|$ of gyrotriangle ABC is given by each of the equations*

$$|ABC| = \frac{2}{1 + \gamma_a + \gamma_b + \gamma_c} \gamma_a a \gamma_{h_a} h_a$$

$$|ABC| = \frac{2}{1 + \gamma_a + \gamma_b + \gamma_c} \gamma_b b \gamma_{h_b} h_b \tag{4.148}$$

$$|ABC| = \frac{2}{1 + \gamma_a + \gamma_b + \gamma_c} \gamma_c c \gamma_{h_c} h_c \,,$$

*where $h_c = h$ is the gyroaltitude of gyrotriangle ABC drawn from side **c**, as shown in Fig. 4.10, and where h_a and h_b are the gyroaltitudes of gyrotriangle ABC drawn from sides **a** and **b**, respectively.*

Proof. The proof of the third equation in (4.148), where $h_c = h$, follows immediately by substituting $s^2 \tan \frac{\delta}{2}$ from (4.145) into the gyrotriangle gyroarea definition (4.147). The proof of the first and second equations in (4.148) follows from the third equation in (4.148) and (4.146). □

In the Newtonian-Euclidean limit of large s, $s \to \infty$, gamma factors tend to 1, so that the gyrotriangle gyroarea formula (4.148) reduces to the common Euclidean triangle area formula. Along with the analogy that the gyrotriangle gyroarea formula (4.148) shares with its Euclidean counterpart there is a remarkable disanalogy. While two triangles with equal bases and equal altitudes have equal areas, two gyrotriangles with equal bases and equal gyroaltitudes need not have equal gyroareas.

4.17 GYROTRIANGLE SIMILARITY

Definition 4.30. (Euclidean Triangle Similarity). Two triangles in a Euclidean vector space are similar if two angles of one triangle are congruent to two angles of the other triangle.

If two triangles are similar, then all the three angles of one triangle are congruent to all the three angles of the other triangle. Hence the relation of triangle similarity is transitive, that is, if triangle T_1 is similar to triangle T_2, and if triangle T_2 is similar to triangle T_3, then triangle T_1 is similar to triangle T_3.

Definition 4.30 suggests the following analogous definition of gyrotriangle similarity.

Definition 4.31. (Hyperbolic Gyrotriangle Similarity). Two gyrotriangles in a gyrovector space are similar if two gyroangles of one gyrotriangle are congruent to two gyroangles of the other gyrotriangle.

Unlike triangle similarity, gyrotriangle similarity is not transitive. Yet, gyrotriangle similarity shares a remarkable analogy with triangle similarity, stated in Theorem 4.32 below.

Theorem 4.32. *Let $A_k B_k C_k$, $k = 1, 2$, be two similar gyrotriangles in an Einstein gyrovector space that, in the standard gyrotriangle notation in Fig. 4.4, p. 117, have respectively, side gyrolengths a_k, b_k, c_k, and gyroangles α_k, β_k and γ_k with*

$$\alpha_1 = \alpha_2 =: \alpha$$
$$\beta_1 = \beta_2 =: \beta. \tag{4.149}$$

Then,

$$\frac{\gamma_{a_1} a_1}{\gamma_{b_1} b_1} = \frac{\gamma_{a_2} a_2}{\gamma_{b_2} b_2}, \tag{4.150}$$

where a_k and b_k are the gyrolengths of the sides opposite to gyroangles α and β, respectively, in gyrotriangles $A_k B_k C_k$, $k = 1, 2$.

Proof. Let us introduce to each of the two gyrotriangles $A_k B_k C_k$, $k = 1, 2$, its altitude $\mathbf{h}_k$ drawn from vertex C_k, as partially shown in Fig. 4.10, and let $h_k = \|\mathbf{h}_k\|$. Then, by elementary gyrotrigonometry, shown in Fig. 4.6, p. 136, we have

$$\sin \alpha_k = \frac{\gamma_{h_k} h_k}{\gamma_{b_k} b_k}$$

$$\sin \beta_k = \frac{\gamma_{h_k} h_k}{\gamma_{a_k} a_k} \, .$$

(4.151)

But, $\alpha_1 = \alpha_2$ and $\beta_1 = \beta_2$. Hence, it follows from (4.151) that

$$\frac{\gamma_{h_1} h_1}{\gamma_{b_1} b_1} = \frac{\gamma_{h_2} h_2}{\gamma_{b_2} b_2}$$

$$\frac{\gamma_{h_1} h_1}{\gamma_{a_1} a_1} = \frac{\gamma_{h_2} h_2}{\gamma_{a_2} a_2} \, ,$$

(4.152)

implying

$$\frac{\gamma_{a_1} a_1}{\gamma_{b_1} b_1} = \frac{\gamma_{a_2} a_2}{\gamma_{b_2} b_2} \, ,$$

(4.153)

as desired. $\square$

4.18   THE GYROANGLE BISECTOR THEOREM

Theorem 4.33.   (The Gyrotriangle Bisector Theorem). *Let $A B_1 B_2$ be a gyrotriangle in an Einstein gyrovector space $(\mathbb{V}_s, \oplus, \otimes)$, and let P be a point lying on side $B_1 B_2$ of the gyrotriangle such that AP is a bisector of gyroangle $\angle B_1 A B_2$, as shown in Fig. 4.11, with $\alpha_1 = \alpha_2$, for an Einstein gyrovector plane $\mathbb{V}_s = \mathbb{R}_s^2$. Then, in the notation of Fig. 4.11,*

$$\frac{\gamma_{a_1} a_1}{\gamma_{a_2} a_2} = \frac{\gamma_{b_1} b_1}{\gamma_{b_2} b_2} \, .$$

(4.154)

Theorem 4.33 above is a special case of Theorem 4.34 below, corresponding to $\alpha_1 = \alpha_2$. The proof of Theorem 4.33 is therefore included in the proof of the following Theorem 4.34, for the special case when $\alpha_1 = \alpha_2$.

Theorem 4.34.   (The Generalized Gyrotriangle Bisector Theorem). *Let $A B_1 B_2$ be a gyrotriangle in an Einstein gyrovector space $(\mathbb{V}_s, \oplus, \otimes)$, and let P be a point lying on side $B_1 B_2$ of the gyrotriangle,*

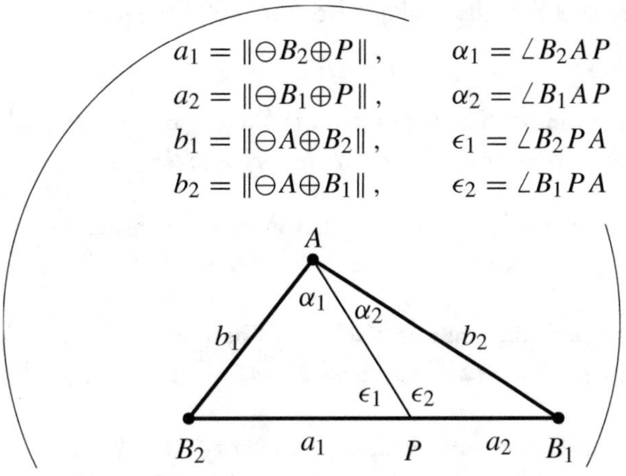

$$a_1 = \|\ominus B_2 \oplus P\|, \qquad \alpha_1 = \angle B_2 A P$$
$$a_2 = \|\ominus B_1 \oplus P\|, \qquad \alpha_2 = \angle B_1 A P$$
$$b_1 = \|\ominus A \oplus B_2\|, \qquad \epsilon_1 = \angle B_2 P A$$
$$b_2 = \|\ominus A \oplus B_1\|, \qquad \epsilon_2 = \angle B_1 P A$$

Figure 4.11: The Generalized Gyroangle Bisector Theorem. Gyrotriangle AB_1B_2 in an Einstein gyrovector plane $(\mathbb{R}_s^2, \oplus, \otimes)$ illustrates the Generalized Gyroangle Bisector Theorem 4.34. In the special case when $\alpha_1 = \alpha_2$ it illustrates the Gyroangle Bisector Theorem 4.33.

as shown in Fig. 4.11 for an Einstein gyrovector plane $\mathbb{V}_s = \mathbb{R}_s^2$. *Then, in the notation of Fig. 4.11,*

$$\frac{\gamma_{a_1} a_1}{\gamma_{a_2} a_2} = \frac{\gamma_{b_1} b_1 \sin \alpha_1}{\gamma_{b_2} b_2 \sin \alpha_2}. \tag{4.155}$$

Proof. With the notation of Fig. 4.11 we have, by the law of gyrosines for each of the two gyrotriangles AB_2P and AB_1P,

$$\frac{\gamma_{a_1} a_1}{\sin \alpha_1} = \frac{\gamma_{b_1} b_1}{\sin \epsilon_1}$$

$$\frac{\gamma_{a_2} a_2}{\sin \alpha_2} = \frac{\gamma_{b_2} b_2}{\sin \epsilon_2}. \tag{4.156}$$

Since $\epsilon_1 + \epsilon_2 = \pi$, we have $\sin \epsilon_1 = \sin \epsilon_2$ so that (4.156) implies the desired result (4.155) of the theorem. □

4.19 THE HYPERBOLIC STEINER – LEHMUS THEOREM

According to A.S. Posamentier [39, p. 88], the proof of the Steiner – Lehmus Theorem in Euclidean geometry is regarded as one of the most difficult in elementary Euclidean geometry. Its long history is described, for instance, in [26] and [21] and references therein. Following the presentation of the Steiner – Lehmus Theorem we present and prove its counterpart in hyperbolic geometry. Our proof is analogous to the one found, for instance, in [39, p. 90] and [45]. In particular, our use of the gyroparallelogram in our proof is fully analogous to the use of the parallelogram in the proof of the classical counterpart.

Theorem 4.35.   (The Euclidean Steiner – Lehmus Theorem). *Any triangle having two equal internal angle bisectors (each measured from a vertex to the opposite side) is isosceles.*

Unlike its trivial converse, Theorem 4.35 has attracted considerable attention, as evidenced from its long history. The hyperbolic counterpart of Theorem 4.35 follows.

Theorem 4.36.   (The Hyperbolic Steiner – Lehmus Theorem). *Any gyrotriangle having two equal internal gyroangle bisectors (each measured from a vertex to the opposite side) is isosceles.*

Proof. Let ABC be a gyrotriangle in an Einstein gyrovector space $(\mathbb{V}_s, \oplus, \otimes)$, shown in Fig. 4.12 for an Einstein gyrovector plane $(\mathbb{R}_s^2, \oplus, \otimes)$, and let BB' and CC' be the respective internal gyroangle bisectors of gyroangles $2\beta = \angle ABC$ and $2\gamma = \angle ACB$ in gyrotriangle ABC. Assuming that these gyrotriangle bisectors have equal gyrolengths,

$$|BB'| = |CC'|, \qquad (4.157)$$

we have to prove that

$$|AB| = |AC|. \qquad (4.158)$$

Seeking a contradiction, we assume that gyrotriangle ABC is not isosceles. Then, by Theorem 4.12, p. 130, without loss of generality we have $2\beta > 2\gamma$ (otherwise, we interchange α and β) or, equivalently,

$$\beta > \gamma. \qquad (4.159)$$

Then, gyrotriangles $BB'C$ and BCC' possess two congruent sides, $|BB'| = |CC'|$, and the common side, $|BC|$, so that by Lemma 4.14, p. 131,

$$|B'C| > |BC'|. \qquad (4.160)$$

Let us augment gyrotriangle $BB'C'$ into a gyroparallelogram $BB'DC'$ by incorporating the point D given by the gyroparallelogram condition, (3.135), p. 91,

$$D = (B' \boxplus C') \ominus B. \qquad (4.161)$$

By Theorem 4.4, p. 113, for the resulting gyroparallelogram $BB'DC'$, Fig. 4.12, and by (4.157), we have

$$|C'D| = |BB'| = |CC'|$$

$$|BC'| = |B'D| \qquad (4.162)$$

$$\beta = \epsilon,$$

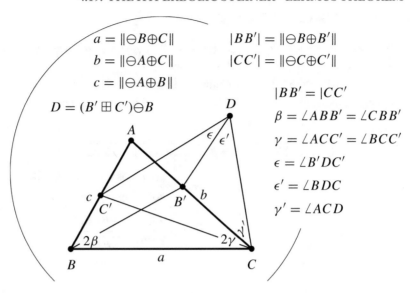

$a = \|\ominus B \oplus C\|$

$b = \|\ominus A \oplus C\|$

$c = \|\ominus A \oplus B\|$

$D = (B' \boxplus C') \ominus B$

$|BB'| = \|\ominus B \oplus B'\|$

$|CC'| = \|\ominus C \oplus C'\|$

$|BB' = |CC'$

$\beta = \angle ABB' = \angle CBB'$

$\gamma = \angle ACC' = \angle BCC'$

$\epsilon = \angle B'DC'$

$\epsilon' = \angle BDC$

$\gamma' = \angle ACD$

Figure 4.12: Illustrating the hyperbolic Steiner – Lehmus Theorem 4.36 in an Einstein gyrovector plane $(\mathbb{R}_s^2, \oplus, \otimes)$.

where $\epsilon = \angle B'DC'$, so that gyrotriangle $CC'D$ is isosceles, $|C'C| = |C'D|$. Hence, by (4.162),

$$\beta + \epsilon' = \epsilon + \epsilon' = \gamma + \gamma', \tag{4.163}$$

where $\epsilon' = \angle B'DC$ and $\gamma' = \angle B'CD$, as shown in Fig. 4.12. Hence, by (4.159) and (4.163),

$$\epsilon' < \gamma'. \tag{4.164}$$

It follows from (4.164), by Lemma 4.14, p. 131, that in gyrotriangle $B'CD$ we have

$$|B'C| < |B'D|. \tag{4.165}$$

Hence, by (4.165) and (4.162),

$$|B'C| < |B'D| = |BC'|, \tag{4.166}$$

thus contradicting (4.160).

The contradiction between (4.160) and (4.166) results from the assumption $\beta > \gamma$ in (4.159). Similarly, also the assumption $\gamma > \beta$ results in a contradiction. Hence $\beta = \gamma$ as desired.   □

4.20   THE URQUHART THEOREM

In order to set the stage for the presentation of its gyro-analog, Urquhart theorem in a Euclidean plane and its proof are presented in this section.

Let us consider the following chain of identities, each of which is valid in trigonometry [20] and, hence, in gyrotrigonometry as well, as explained in Secs. 4.2 and 4.5.

$$
\begin{aligned}
\frac{2}{1 - \tan\alpha \tan\beta} &= \frac{2\cos\alpha \cos\beta}{\cos\alpha \cos\beta - \sin\alpha \sin\beta} \\
&= 1 + \frac{\cos\alpha \cos\beta + \sin\alpha \sin\beta}{\cos\alpha \cos\beta - \sin\alpha \sin\beta} \\
&= 1 + \frac{2\sin(\alpha + \beta)\cos(\alpha - \beta)}{2\sin(\alpha + \beta)\cos(\alpha + \beta)} \\
&= 1 + \frac{\sin 2\alpha + \sin 2\beta}{\sin(2\alpha + 2\beta)} .
\end{aligned}
\tag{4.167}
$$

The chain of identities (4.167) gives rise to the identity

$$
\frac{\sin\alpha + \sin\beta}{\sin(\alpha + \beta)} = -1 + \frac{2}{1 - \tan\frac{\alpha}{2}\tan\frac{\beta}{2}} ,
\tag{4.168}
$$

when $\sin(\alpha + \beta) \neq 0$, that we will employ both trigonometrically, in this section, and gyrotrigono-metrically, in Sec. 4.21.

Lemma 4.37.   (Breusch's Lemma in Euclidean Geometry [50]**).** *Let ABC_k, $k = 1, 2$, be two triangles in a Euclidean plane $\mathbb{R}^2$ with common side AB, with side lengths a_k, b_k, c_k, and with angles α_k, β_k, γ_k, as shown in Fig. 4.13. Then*

$$
a_1 + b_1 = a_2 + b_2 \Longleftrightarrow \tan\frac{\alpha_1}{2}\tan\frac{\beta_1}{2} = \tan\frac{\alpha_2}{2}\tan\frac{\beta_2}{2} ,
\tag{4.169}
$$

or equivalently,

$$
P_e(ABC_1) = P_e(ABC_2) \Longleftrightarrow \tan\frac{\alpha_1}{2}\tan\frac{\beta_1}{2} = \tan\frac{\alpha_2}{2}\tan\frac{\beta_2}{2} ,
\tag{4.170}
$$

where $P_e(ABC_1)$ is the perimeter of triangle ABC_1, given by, Fig. 4.13,

$$
P_e(ABC_1) = a_1 + b_1 + c ,
\tag{4.171}
$$

etc.

$$a_k = |BC_k| = \|-B+C_k\|$$

$$b_k = |AC_k| = \|-A+C_k\|$$

$$c = |AB| = \|-A+B\|$$

$$a_k = |BC_k| = \|\ominus B \oplus C_k\|$$

$$b_k = |AC_k| = \|\ominus A \oplus C_k\|$$

$$c = |AB| = \|\ominus A \oplus B\|$$

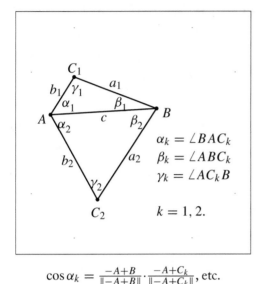

$$\alpha_k = \angle BAC_k$$
$$\beta_k = \angle ABC_k$$
$$\gamma_k = \angle AC_k B$$

$$k = 1, 2.$$

$$\cos \alpha_k = \frac{-A+B}{\|-A+B\|} \cdot \frac{-A+C_k}{\|-A+C_k\|}, \text{etc.}$$

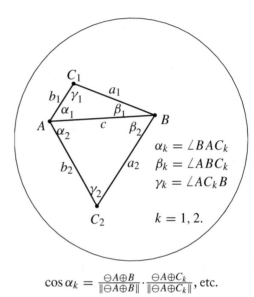

$$\alpha_k = \angle BAC_k$$
$$\beta_k = \angle ABC_k$$
$$\gamma_k = \angle AC_k B$$

$$k = 1, 2.$$

$$\cos \alpha_k = \frac{\ominus A \oplus B}{\|\ominus A \oplus B\|} \cdot \frac{\ominus A \oplus C_k}{\|\ominus A \oplus C_k\|}, \text{etc.}$$

Figure 4.13: Euclidean Plane: Two triangles, ABC_k, $k = 1, 2$, with common side AB in a Euclidean plane $\mathbb{R}^2$ are shown, along with their Euclidean geometric and trigonometric properties, to illustrate Breusch's Lemma 4.37 in Euclidean geometry.

Figure 4.14: Hyperbolic Plane: Two gyrotriangles, ABC_k, $k = 1, 2$, with common side AB in an Einstein gyrovector plane $(\mathbb{R}^2, \oplus, \otimes)$ are shown, along with their gyrogeometric and gyrotrigonometric properties, to illustrate Breusch's Lemma 4.39 in hyperbolic geometry.

Proof. By applying the trigonometric law of sines to triangles ABC_k, $k = 1, 2$, in Fig. 4.13, and employing the trigonometric identity (4.168), we have

$$
\begin{aligned}
\frac{a_k + b_k}{c} &= \frac{\sin \alpha_k + \sin \beta_k}{\sin \gamma_k} \\
&= \frac{\sin \alpha_k + \sin \beta_k}{\sin(\pi - \alpha_k - \beta_k)} \\
&= \frac{\sin \alpha_k + \sin \beta_k}{\sin(\alpha_k + \beta_k)} \\
&= -1 + \frac{2}{1 - \tan \frac{\alpha_k}{2} \tan \frac{\beta_k}{2}},
\end{aligned}
\tag{4.172}
$$

thus verifying (4.169).   □

Lemma 4.37 enables us to prove the following theorem of L.M. Urquhart (1902-1962), who discovered it when considering some of the fundamental concepts of special relativity theory [20, 7].

Theorem 4.38.   (Urquhart's Theorem in Euclidean Geometry [20]**).** *Let* AD_1BD_2 *be a concave quadrilateral in a Euclidean plane* $\mathbb{R}^2$, *and let* AD_1 *meet* D_2B *at* C_1, *and* AD_2 *meet* D_1B *at* C_2, *as shown in Fig. 4.15, p. 159. Then*

$$|AC_1| + |C_1B| = |AC_2| + |C_2B| \iff |AD_1| + |D_1B| = |AD_2| + |D_2B|, \qquad (4.173)$$

or equivalently, but in the alternative notation in Fig. 4.15,

$$a_1 + b_1 = a_2 + b_2 \iff a_1' + b_1' = a_2' + b_2', \qquad (4.174)$$

or equivalently,

$$P_e(ABC_1) = P_e(ABC_2) \iff P_e(ABD_1) = P_e(ABD_2). \qquad (4.175)$$

Proof. To prove the theorem we apply Breusch's Lemma 4.37 for triangles ABC_k, $k = 1, 2$, in Fig. 4.13 to

(i) triangles $ABC_k, k = 1, 2$, in Fig. 4.15, and to

(ii) triangles $ABD_k, k = 1, 2$, in the same Fig. 4.15.

Applying Lemma 4.37 to triangles $ABC_k, k = 1, 2$, in Fig. 4.15, we have

$$|AC_1| + |C_1B| = |AC_2| + |C_2B| \iff \tan \tfrac{\alpha_1}{2} \tan \tfrac{\beta_1}{2} = \tan \tfrac{\alpha_2}{2} \tan \tfrac{\beta_2}{2}. \qquad (4.176)$$

Note that, within the notation of Fig. 4.15, the assertions in (4.176) and in (4.169) are identically the same.

In a similar way we now apply Lemma 4.37 to triangles $ABD_k, k = 1, 2$, in Fig. 4.15, obtaining the following chain of implications:

$$|AD_1| + |D_1B| = |AD_2| + |D_2B| \iff \tan \tfrac{\alpha_1}{2} \tan(\tfrac{\pi}{2} - \tfrac{\beta_2}{2}) = \tan \tfrac{\alpha_2}{2} \tan(\tfrac{\pi}{2} - \tfrac{\beta_1}{2})$$

$$\iff \tan \tfrac{\alpha_1}{2} \cot \tfrac{\beta_2}{2} = \tan \tfrac{\alpha_2}{2} \cot \tfrac{\beta_1}{2} \qquad (4.177)$$

$$\iff \tan \tfrac{\alpha_1}{2} \tan \tfrac{\beta_1}{2} = \tan \tfrac{\alpha_2}{2} \tan \tfrac{\beta_2}{2}.$$

The result (4.173) of the theorem follows straightforwardly from the implications in (4.176) and (4.177).

Finally, the equivalence between (4.173) and (4.174) is a matter of notation, and the equivalence between (4.174) and (4.175) is immediate.   □

4.21 THE HYPERBOLIC URQUHART THEOREM

Both Breusch's lemma and Urquhart's theorem in a Euclidean plane admit natural counterparts in an Einstein gyrovector plane, as shown in this section.

Lemma 4.39.   **(Breusch's Lemma in Hyperbolic Geometry)** *Let ABC_k, $k = 1, 2$, be two gyrotriangles in an Einstein gyrovector plane $(\mathbb{R}^2_s, \oplus, \otimes)$ with common side AB, with side gyrolengths a_k, b_k, c_k, and with gyroangles α_k, β_k, γ_k, as shown in Fig. 4.14, p. 155. Then*

$$\frac{\gamma_{a_1} a_1 + \gamma_{b_1} b_1}{\gamma_{a_1} + \gamma_{b_1}} = \frac{\gamma_{a_2} a_2 + \gamma_{b_2} b_2}{\gamma_{a_2} + \gamma_{b_2}} \iff \tan \tfrac{\alpha_1}{2} \tan \tfrac{\beta_1}{2} = \tan \tfrac{\alpha_2}{2} \tan \tfrac{\beta_2}{2}, \tag{4.178}$$

or equivalently,

$$a_1 \oplus b_1 = a_2 \oplus b_2 \iff \tan \tfrac{\alpha_1}{2} \tan \tfrac{\beta_1}{2} = \tan \tfrac{\alpha_2}{2} \tan \tfrac{\beta_2}{2}, \tag{4.179}$$

or equivalently,

$$P_h(ABC_1) = P_h(ABC_2) \iff \tan \tfrac{\alpha_1}{2} \tan \tfrac{\beta_1}{2} = \tan \tfrac{\alpha_2}{2} \tan \tfrac{\beta_2}{2}, \tag{4.180}$$

where $P_h(ABC_1)$ is the gyroperimeter of gyrotriangle ABC_1, given by, Fig. 4.14,

$$P_e(ABC_1) = a_1 \oplus b_1 \oplus c, \tag{4.181}$$

etc.

Proof. Applying the gyrotriangle identity (4.50), p. 127, to each of the two gyrotriangles ABC_k, $k = 1, 2$, in Fig. 4.14, and employing the gyrotrigonometric identity (4.168), we have

$$\frac{1}{s} \sqrt{\frac{\gamma_c + 1}{\gamma_c - 1}} \frac{\gamma_{a_k} a_k + \gamma_{b_k} b_k}{\gamma_{a_k} + \gamma_{b_k}} = \frac{\sin \alpha_k + \sin \beta_k}{\sin(\alpha_k + \beta_k)} = -1 + \frac{2}{1 - \tan \tfrac{\alpha_k}{2} \tan \tfrac{\beta_k}{2}}, \tag{4.182}$$

$k = 1, 2$, where $c = |AB|$.

Hence, the extreme left-hand side of (4.182) is independent of whether $k = 1$ or $k = 2$ if and only if the extreme right-hand side of (4.182) is independent of whether $k = 1$ or $k = 2$, thus verifying the result (4.178) of the Lemma.

The equation on the left-hand side of implication (4.178) can be written as

$$\tfrac{1}{2} \otimes (a_1 \boxplus b_1) = \tfrac{1}{2} \otimes (a_2 \boxplus b_2) \tag{4.183}$$

according to (2.89), p. 49. However, (4.183) is equivalent to

$$a_1 \boxplus b_1 = a_2 \boxplus b_2, \tag{4.184}$$

which, in turn, is equivalent to

$$a_1 \oplus b_1 = a_2 \oplus b_2 \tag{4.185}$$

according to (2.77), p. 46. The latter is equivalent to

$$P_h(ABC_1) := a_1 \oplus b_1 \oplus c = a_2 \oplus b_2 \oplus c =: P_h(ABC_2), \tag{4.186}$$

thus verifying (4.179) and (4.180) of the Lemma.   □

Unlike implication (4.178), which is peculiar to Einstein gyrovector spaces, the implications in (4.179) and (4.180) are model indepenent so that, in particular, they are valid in Möbius gyrovector spaces as well.

With the help of the Hyperbolic Breusch Lemma 4.39 we prove the following Hyperbolic Urquhart's Theorem (4.40), illustrated in Figs. 4.15 – 4.16.

Theorem 4.40.   (Urquhart's Theorem in Hyperbolic Geometry). *Let AD_1BD_2 be a concave gyro-quadrilateral in an Einstein gyrovector plane $(\mathbb{R}_s^2, \oplus, \otimes)$, and let AD_1 meet D_2B at C_1, and AD_2 meet D_1B at C_2, as shown in Fig. 4.16. Then, in the notation of Fig. 4.16,*

$$\frac{\gamma_{a_1} a_1 + \gamma_{b_1} b_1}{\gamma_{a_1} + \gamma_{b_1}} = \frac{\gamma_{a_2} a_2 + \gamma_{b_2} b_2}{\gamma_{a_2} + \gamma_{b_2}} \Longleftrightarrow \frac{\gamma_{a'_1} a'_1 + \gamma_{b'_1} b'_1}{\gamma_{a'_1} + \gamma_{b'_1}} = \frac{\gamma_{a'_2} a'_2 + \gamma_{b'_2} b'_2}{\gamma_{a'_2} + \gamma_{b'_2}}, \tag{4.187}$$

or equivalently,

$$a_1 \oplus b_1 = a_2 \oplus b_2 \Longleftrightarrow a'_1 \oplus b'_1 = a'_2 \oplus b'_2, \tag{4.188}$$

or equivalently,

$$P_h(ABC_1) = P_h(ABC_2) \Longleftrightarrow P_h(ABD_1) = P_h(ABD_2). \tag{4.189}$$

Proof. To prove the theorem we apply Lemma 4.39 for gyrotriangles $ABC_k, k = 1, 2$, in Fig. 4.14 to

  (i) gyrotriangles $ABC_k, k = 1, 2$, in Fig. 4.16, and to

 (ii) gyrotriangles $ABD_k, k = 1, 2$, in the same Fig. 4.16.

Applying Lemma 4.39 to gyrotriangles $ABC_k, k = 1, 2$, in Fig. 4.16, we have

$$\frac{\gamma_{a_1} a_1 + \gamma_{b_1} b_1}{\gamma_{a_1} + \gamma_{b_1}} = \frac{\gamma_{a_2} a_2 + \gamma_{b_2} b_2}{\gamma_{a_2} + \gamma_{b_2}} \Longleftrightarrow \tan \tfrac{\alpha_1}{2} \tan \tfrac{\beta_1}{2} = \tan \tfrac{\alpha_2}{2} \tan \tfrac{\beta_2}{2}. \tag{4.190}$$

Note that (4.190) and in (4.178) are identical to each other.

$$a_k = |BC_k| = \| - B + C_k \|$$

$$b_k = |AC_k| = \| - A + C_k \|$$

$$c = |AB| = \| - A + B \|$$

$$a_k' = |BD_k| = \| - B + D_k \|$$

$$b_k' = |AD_k| = \| - A + D_k \|$$

$$a_k = |BC_k| = \| \ominus B \oplus C_k \|$$

$$b_k = |AC_k| = \| \ominus A \oplus C_k \|$$

$$c = |AB| = \| \ominus A \oplus B \|$$

$$a_k' = |BD_k| = \| \ominus B \oplus D_k \|$$

$$b_k' = |AD_k| = \| \ominus A \oplus D_k \|$$

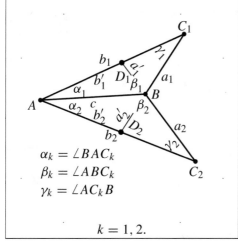

$$\alpha_k = \angle BAC_k$$
$$\beta_k = \angle ABC_k$$
$$\gamma_k = \angle AC_k B$$

$$k = 1, 2.$$

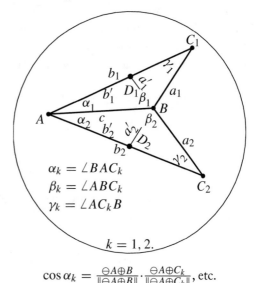

$$\alpha_k = \angle BAC_k$$
$$\beta_k = \angle ABC_k$$
$$\gamma_k = \angle AC_k B$$

$$k = 1, 2.$$

$$\cos \alpha_k = \frac{-A+B}{\| -A+B \|} \cdot \frac{-A+C_k}{\| -A+C_k \|}, \text{etc.}$$

$$\angle ABD_1 = \pi - \beta_2, \quad \angle ABD_2 = \pi - \beta_1$$

$$\cos \alpha_k = \frac{\ominus A \oplus B}{\| \ominus A \oplus B \|} \cdot \frac{\ominus A \oplus C_k}{\| \ominus A \oplus C_k \|}, \text{etc.}$$

$$\angle ABD_1 = \pi - \beta_2, \quad \angle ABD_2 = \pi - \beta_1$$

Figure 4.15: Euclidean Plane: Two triangles, ABC_k, $k = 1, 2$, with common side AB in a Euclidean plane $\mathbb{R}^2$ are shown, along with their Euclidean geometric properties, to illustrate Urquhart's Theorem 4.38, p. 156, in Euclidean geometry.

Figure 4.16: Hyperbolic Plane: Two gyrotriangles, $ABC_k, k = 1, 2$, with common side AB in an Einstein gyrovector plane $(\mathbb{R}^2, \oplus, \otimes)$ are shown, along with their gyrogeometric properties, to illustrate Urquhart's Theorem 4.40, p. 158, in hyperbolic geometry.

In a similar way we now apply Lemma 4.39 to gyrotriangles $ABD_k, k = 1, 2$, in Fig. 4.16, obtaining the following chain of implications:

$$\frac{\gamma_{a_1'} a_1' + \gamma_{b_1'} b_1'}{\gamma_{a_1'} + \gamma_{b_1'}} = \frac{\gamma_{a_2'} a_2' + \gamma_{b_2'} b_2'}{\gamma_{a_2'} + \gamma_{b_2'}} \iff \tan \frac{\alpha_1}{2} \tan(\frac{\pi}{2} - \frac{\beta_2}{2}) = \tan \frac{\alpha_2}{2} \tan(\frac{\pi}{2} - \frac{\beta_1}{2})$$

$$\iff \tan \frac{\alpha_1}{2} \cot \frac{\beta_2}{2} = \tan \frac{\alpha_2}{2} \cot \frac{\beta_1}{2}$$

$$\iff \tan \frac{\alpha_1}{2} \tan \frac{\beta_1}{2} = \tan \frac{\alpha_2}{2} \tan \frac{\beta_2}{2}.$$

(4.191)

The result (4.187) of the theorem follows straightforwardly from the implications in (4.190) and (4.191).

Finally, the proof of the equivalence between (4.188) and (4.187) and between (4.189) and (4.188) is obtained in terms of (4.183)–(4.185) as in the proof of Lemma 4.39.   □

Unlike implication (4.187), which is peculiar to Einstein gyrovector spaces, the implications in (4.188) and (4.189) are model indepenent so that, in particular, they are valid in Möbius gyrovector spaces as well.

The analogies that the results (4.174)–(4.175) share with the results (4.188)–(4.189) are remarkable, demonstrating once again the role of the gamma factor in capturing analogies between Euclidean geometry and the Beltrami-Klein model of hyperbolic geometry, along with well-known analogies between classical and relativistic mechanics.

4.22   THE GYROPARALLELOGRAM GYROANGLES

Let

$$
\alpha = \angle BAB' = \angle B'A'B
$$
$$
\beta = \angle A'BA = \angle AB'A'
$$

(4.192)

be the two distinct gyroangles of the gyroparallelogram $ABA'B'$ in Fig. 4.17. They are related to each other by the equations

$$
\cos\alpha = \frac{\sqrt{\gamma_a^2 - 1}\sqrt{\gamma_b^2 - 1} - (1 + \gamma_a\gamma_b)\cos\beta}{1 + \gamma_a\gamma_b - \sqrt{\gamma_a^2 - 1}\sqrt{\gamma_b^2 - 1}\cos\beta}
$$

(4.193)

$$
\cos\beta = \frac{\sqrt{\gamma_a^2 - 1}\sqrt{\gamma_b^2 - 1} - (1 + \gamma_a\gamma_b)\cos\alpha}{1 + \gamma_a\gamma_b - \sqrt{\gamma_a^2 - 1}\sqrt{\gamma_b^2 - 1}\cos\alpha},
$$

as one can see by translating [55, Theorem 8.59, p. 297] from Möbius to Einstein gyrovector spaces by means of [55, Eqs. (6.309)–(6.310), p. 208].

It follows from (4.193) that

$$
\sin\alpha = \frac{\gamma_a + \gamma_b}{1 + \gamma_a\gamma_b - \sqrt{\gamma_a^2 - 1}\sqrt{\gamma_b^2 - 1}\cos\beta}\sin\beta
$$

(4.194)

$$
\sin\beta = \frac{\gamma_a + \gamma_b}{1 + \gamma_a\gamma_b - \sqrt{\gamma_a^2 - 1}\sqrt{\gamma_b^2 - 1}\cos\alpha}\sin\alpha .
$$

In the Newtonian limit, $s \to \infty$, γ_a and γ_b reduce to 1. Hence, Identities (4.193)–(4.194) for the gyroparallelogram reduce to the identities $\cos\alpha = -\cos\beta$ and $\sin\alpha = \sin\beta$ for the parallelogram, implying $\beta = \pi - \alpha$ as expected for parallelograms in Euclidean geometry.

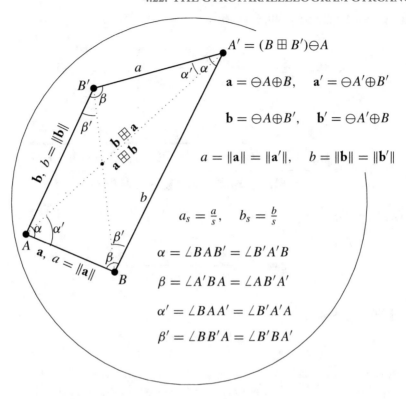

$$A' = (B \boxplus B') \ominus A$$

$$\mathbf{a} = \ominus A \oplus B, \quad \mathbf{a}' = \ominus A' \oplus B'$$

$$\mathbf{b} = \ominus A \oplus B', \quad \mathbf{b}' = \ominus A' \oplus B$$

$$a = \|\mathbf{a}\| = \|\mathbf{a}'\|, \quad b = \|\mathbf{b}\| = \|\mathbf{b}'\|$$

$$a_s = \frac{a}{s}, \quad b_s = \frac{b}{s}$$

$$\alpha = \angle BAB' = \angle B'A'B$$

$$\beta = \angle A'BA = \angle AB'A'$$

$$\alpha' = \angle BAA' = \angle B'A'A$$

$$\beta' = \angle BB'A = \angle B'BA'$$

Figure 4.17: The Gyroangles of the Einstein Gyroparallelogram in Einstein gyrovector spaces $(\mathbb{V}_s, \oplus, \otimes)$. The two distinct gyroangles α and β of a gyroparallelogram in an Einstein gyrovector space are related to each other by (4.193)–(4.194).

The gyroparallelogram addition law, (3.139), p. 93, is shown graphically in Fig. 3.10, p. 98, of the Einstein gyroparallelogram and in Fig. 3.11, p. 99, of the Möbius gyroparallelogram. Accordingly, the gyrodiagonal AA' of a gyroparallelogram $ABA'B'$ in Fig. 4.17 satisfies the gyrovector equation

$$\ominus A \oplus A' = \mathbf{a} \boxplus \mathbf{b} = \mathbf{b} \boxplus \mathbf{a} \tag{4.195}$$

where $\mathbf{a}$ and $\mathbf{b}$ are the gyrovectors

$$\begin{aligned} \mathbf{a} &= \ominus A \oplus B \\ \mathbf{b} &= \ominus A \oplus B', \end{aligned} \tag{4.196}$$

with magnitudes

$$\begin{aligned} a &= \|\mathbf{a}\| \\ b &= \|\mathbf{b}\|. \end{aligned} \tag{4.197}$$

Furthermore, by (2.89), p. 49, we have

$$\mathbf{a} \boxplus \mathbf{b} = \frac{\gamma_{\mathbf{a}} + \gamma_{\mathbf{b}}}{\gamma_{\mathbf{a}}^2 + \gamma_{\mathbf{b}}^2 + \gamma_{\mathbf{a}} \gamma_{\mathbf{b}} (1 + \mathbf{a}_s \cdot \mathbf{b}_s) - 1} (\gamma_{\mathbf{a}} \mathbf{a} + \gamma_{\mathbf{b}} \mathbf{b}), \tag{4.198}$$

implying

$$\frac{1}{s} \|\mathbf{a} \boxplus \mathbf{b}\| = \frac{(\gamma_{\mathbf{a}} + \gamma_{\mathbf{b}}) \sqrt{(\gamma_{\mathbf{a}} + \gamma_{\mathbf{b}})^2 - 2\{\gamma_{\mathbf{a}} \gamma_{\mathbf{b}} (1 - \mathbf{a}_s \cdot \mathbf{b}_s) + 1\}}}{(\gamma_{\mathbf{a}} + \gamma_{\mathbf{b}})^2 - \gamma_{\mathbf{a}} \gamma_{\mathbf{b}} (1 - \mathbf{a}_s \cdot \mathbf{b}_s) - 1}, \tag{4.199}$$

and, as in (2.91), p. 49,

$$\gamma_{\mathbf{a} \boxplus \mathbf{b}} = \frac{\gamma_{\mathbf{a}}^2 + \gamma_{\mathbf{b}}^2 + \gamma_{\mathbf{a}} \gamma_{\mathbf{b}} (1 + \mathbf{a}_s \cdot \mathbf{b}_s) - 1}{\gamma_{\mathbf{a}} \gamma_{\mathbf{b}} (1 - \mathbf{a}_s \cdot \mathbf{b}_s) + 1}, \tag{4.200}$$

where, as we see from Fig. 4.17,

$$\mathbf{a}_s \cdot \mathbf{b}_s = a_s b_s \cos \alpha. \tag{4.201}$$

Here $\mathbf{a}_s = \mathbf{a}/s$, $a_s = a/s$, etc.

As an application of gyrotrigonometry we calculate below the gyroparallelogram gyroangles α' and β' formed, respectively, by the gyroparallelogram gyrodiagonals AA' and BB', Fig. 4.17, in terms of the gyroparallelogram gyroangles α and β and its side gyrolengths $a = \|\mathbf{a}\|$ and $b = \|\mathbf{b}\|$. We therefore extend the gyrotriangle ABA' of the gyroparallelogram in Fig. 4.17 into a right gyroangled gyrotriangle ACA' in Fig. 4.18; and apply the Einstein gyrotrigonometry, Fig. 4.6, to the resulting two right gyroangled gyrotriangles ACA' and BCA' in Fig. 4.18.

Applying Einstein gyrotrigonometry, Fig. 4.6, p. 136, to the right gyroangled gyrotriangle BCA' in Fig. 4.18 we have

$$\cos(\pi - \beta) = \frac{\|\ominus B \oplus C\|}{b}$$

$$\sin(\pi - \beta) = \frac{\gamma_{\|\ominus A' \oplus C\|} \|\ominus A' \oplus C\|}{\gamma_b b}. \tag{4.202}$$

Hence, noting that $\cos(\pi - \beta) = -\cos \beta$ and $\sin(\pi - \beta) = \sin \beta$ we have

$$\|\ominus B \oplus C\| = -b \cos \beta$$

$$\gamma_{\|\ominus A' \oplus C\|} \|\ominus A' \oplus C\| = \gamma_b b \sin \beta. \tag{4.203}$$

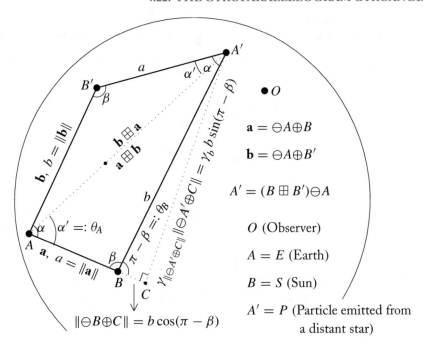

Figure 4.18: Internal gyroangles of a gyroparallelogram in an Einstein gyrovector space $(\mathbb{V}_s, \oplus, \otimes)$. The two distinct gyroangles α and β of a gyroparallelogram in an Einstein gyrovector space are related to each other by (4.193)–(4.194). If B lies between A and C, as shown here, then $\|\ominus A \oplus C\| = \|\ominus A \oplus B\| \oplus \|\ominus B \oplus C\|$, and $\|\ominus B \oplus C\| = b\cos(\pi - \beta)$. If C lies between A and B then $\|\ominus A \oplus C\| = \|\ominus A \oplus B\| \ominus \|\ominus B \oplus C\|$, and $\|\ominus B \oplus C\| = b\cos\beta$.

A relativistic mechanical interpretation of the gyroparallelogram internal gyroangles in this figure is presented in Sec. 4.23, where points of this figure are considered as points of the Einstein velocity space $(\mathbb{R}_s^3, \oplus, \otimes)$ of all relativistically admissible velocities. In particular, (i) point A, $A = E$, represents the velocity of the Earth relative to an observer O at rest relative to the arbitrarily selected point O in $\mathbb{R}_s^3$; (ii) point B, $B = S$, represents the velocity of the Sun relative to the observer O; (iii) point A', $A' = P$, represents the velocity of a particle emitted from a distant star (as, for instance, a photon), relative to the observer O.

It follows from the first equation in (4.203) that the gyrolength of gyrosegment AC, Fig. 4.18, is given by

$$
\begin{aligned}
\|\ominus A \oplus C\| &= \|\ominus A \oplus B\| \oplus \|\ominus B \oplus C\| \\
&= a \oplus (-b\cos\beta) \\
&= a \ominus b\cos\beta \\
&= \frac{a - b\cos\beta}{1 - a_s b_s \cos\beta},
\end{aligned}
\tag{4.204}
$$

where we use the notation in (4.196)–(4.197), and employ Einstein addition for parallel velocities, (2.76), p. 46. The first equation in (4.204) is the gyrotriangle equality (3.82), p. 74.

Applying Einstein gyrotrigonometry, Fig. 4.6, p. 136, to the right gyroangled gyrotriangle ACA' in Fig. 4.18 we have

$$\cos \alpha' = \frac{\|\ominus A \oplus C\|}{\|\mathbf{a} \boxplus \mathbf{b}\|}$$

$$\sin \alpha' = \frac{\gamma_{\ominus A' \oplus C} \|\ominus A' \oplus C\|}{\gamma_{\mathbf{a} \boxplus \mathbf{b}} \|\mathbf{a} \boxplus \mathbf{b}\|} ,$$

(4.205)

so that

$$\cot \alpha' = \frac{\gamma_{\mathbf{a} \boxplus \mathbf{b}} \|\ominus A \oplus C\|}{\gamma_{\ominus A' \oplus C} \|\ominus A' \oplus C\|} .$$

(4.206)

Substituting into (4.206) (i) $\gamma_{\mathbf{a} \boxplus \mathbf{b}}$ from (4.200); (ii) $\|\ominus A \oplus C\|$ from (4.204); and (iii) $\gamma_{\|\ominus A' \oplus C\|} \|\ominus A' \oplus C\|$ from (4.203); and (iv) expressing $\cos \beta$ and $\sin \beta$ in terms of $\cos \alpha$ and $\sin \alpha$ by (4.193)–(4.194), we obtain the following remarkably simple and elegant expression

$$\cot \alpha' = \frac{\gamma_a \, a + \gamma_b \, b \cos \alpha}{\gamma_b \, b \sin \alpha} .$$

(4.207)

Similarly, by symmetry considerations,

$$\cot(\alpha - \alpha') = \frac{\gamma_b \, b + \gamma_a \, a \cos \alpha}{\gamma_a \, a \sin \alpha} .$$

(4.208)

Furthermore, similarly to (4.207)–(4.208) we have

$$\cot \beta' = \frac{\gamma_b \, b + \gamma_a \, a \cos \beta}{\gamma_a \, a \sin \beta}$$

(4.209)

and

$$\cot(\beta - \beta') = \frac{\gamma_a \, a + \gamma_b \, b \cos \beta}{\gamma_b \, b \sin \beta} .$$

(4.210)

To express α' in terms of β we substitute $\cos \alpha$ and $\sin \alpha$ from (4.193)–(4.194) into (4.207) obtaining

$$\cot \alpha' = \gamma_a \frac{a - b \cos \beta}{b \sin \beta} .$$

(4.211)

Using the notation

$$\theta_A = \alpha'$$
$$\theta_B = \pi - \beta ,$$

(4.212)

suggested in Fig. 4.18, Identity (4.211) takes the form

$$\cot \theta_A = \gamma_a \, \frac{a_s + b_s \cos \theta_B}{b_s \sin \theta_B} = \gamma_a \, \frac{\frac{a}{b} + \cos \theta_B}{\sin \theta_B} \,. \tag{4.213}$$

Owing to its relativistic mechanical interpretation, described in Sec. 4.23, Identity (4.213) is called a *relativistic particle aberration formula*, leading to the interpretation of the well-known relativistic effect called *stellar aberration*.

4.23   RELATIVISTIC MECHANICAL INTERPRETATION

In order to demonstrate the intimate connection between the gyrotrigonometry of Einstein gyrovector spaces and relativistic mechanics, we present in this section a relativistic mechanical interpretation of Identity (4.213) in terms of the cosmological effect known as *stellar aberration* [48]. The Einstein three-dimensional gyrovector space $(\mathbb{R}_s^3, \oplus, \otimes)$ represents the velocity space of all uniform relativistically admissible velocities in the Universe. Let O be an arbitrarily selected point, called "origin", of an Einstein three-dimensional gyrovector space $(\mathbb{R}^3, \oplus, \otimes)$, shown in Fig. 4.18 for an Einstein gyrovector plane. This origin of the gyrovector space represents an *Observer*. Furthermore, let the points A, B and A' in Fig. 4.18 represent, respectively, the velocity relative to observer O of the Earth (A), the Sun (B) and a Particle (A') emitted from a distant star. Accordingly, the observer O who is located somewhere in the Universe, sees the Earth and the Sun moving relative to himself with velocities A and B, respectively. Obviously, these two velocities are observer dependent since different observers need not measure the same velocities of the Earth and the Sun relative to themselves. Observer O finds that the (momentarily uniform) velocity of the Sun relative to the Earth is $\mathbf{a} = \ominus A \oplus B$ and, hence, that the speed of the Sun relative to the Earth is $a = \|\ominus A \oplus B\|$. Accordingly:

(*i*) on the one hand, owing to the presence of a gyration (that is, owing to the presence of the relativistic effect known as Thomas precession), the velocity $\mathbf{a} = \ominus A \oplus B$, of the Sun relative to the Earth is observer dependent (that is, it is left gyrotranslation dependent), as we see from the gyrogroup identity (4.19), p. 118, where points A and B are left gyrotranslated by C. But,

(*ii*) on the other hand, the speed $a = \|\ominus A \oplus B\|$ of the Sun relative to the Earth is observer independent (that is, it is left gyrotranslation independent), as we see from (4.24), p. 119.

Each of the two observers,

(*i*) observer $A = E$, who is at rest relative to the Earth, shown as point A in Fig. 4.18, and

(*ii*) observer $B = S$, who is at rest relative to the Sun, shown as point B in Fig. 4.18,

measures the speed and the direction of the velocity P relative to O of a particle, emitted from a distant star. The particle is at rest relative to point $A' = P$ of the Einstein velocity space $(\mathbb{R}_s^3, \oplus, \otimes)$, shown in Fig. 4.18.

(*i*) Observer $E = A$ finds that the velocity of the particle $P = A'$ relative to himself (and, hence, relative to the Earth) is $\ominus A \oplus A' = \ominus E \oplus P = \mathbf{a} \boxplus \mathbf{b}$ in a direction that forms a gyroangle $\theta_A = \angle A'AB$ with the velocity between the Earth and the Sun, as shown in Fig. 4.18. We should note here that for observer $E = A$, who is at rest relative to the Earth, the hyperbolic gyroangle θ_A coincides with its Euclidean counterpart, the angle θ_A, as explained in Fig. 4.1, p. 112.

(*ii*) Similarly, observer $S = B$ finds that the velocity of the particle $P = A'$ relative to himself (and, hence, relative to the Sun) is $\ominus B \oplus A' = \ominus S \oplus P$ in a direction that forms a gyroangle $\theta_B = \angle A'BC$ with the velocity between the Earth and the Sun, as shown in Fig. 4.18. We should note here, again, that for observer $S = B$, who is at rest relative to the Sun, the hyperbolic gyroangle θ_B coincides with its Euclidean counterpart, the angle θ_B, as explained in Fig. 4.1, p. 112. In this sense, gyroangles become angles when measured by appropriate observers.

Owing to the nonvanishing relative velocity, $\mathbf{a} = \ominus A \oplus B$, between the Earth and the Sun, the two gyroangles θ_B and θ_A in Fig. 4.18 are distinct. The effect of the resulting difference $\theta_A - \theta_B$ in the apparent direction of the moving particle $P = A'$ caused by the relative motion between the two observers $E = A$ and $S = B$ is called *particle aberration*, and any relationship between θ_B and θ_A is called a *particle aberration formula*. In the special case when the moving particle is a photon emitted from a star, we use the terms *stellar aberration* and *stellar aberration formula*. Indeed, stellar aberration is an astronomical effect caused by the relative motion between observers on the Earth as the Earth moves through its orbit around the Sun.

Following its relativistic mechanical interpretation, the gyroangle $\theta_A - \theta_B$ is called the aberration gyroangle of the gyroparallelogram $ABA'B'$ in Fig. 4.18.

In the limit when the speed b of the particle approaches the speed of light s, b reduces to s, $b = s$, and (4.213) reduces to the equation,

$$\cot \theta_A = \gamma_a \frac{a_s + \cos \theta_B}{\sin \theta_B} \tag{4.214}$$

that gives rise to the *relativistic photon aberration formula*, known as a *stellar aberration formula*. It implies the following equivalent formulas,

$$\sin \theta_A = \frac{1}{\gamma_a} \frac{\sin \theta_B}{1 + a_s \cos \theta_B}$$

$$\cos \theta_A = \frac{a_s + \cos \theta_B}{1 + a_s \cos \theta_B}. \tag{4.215}$$

The latter can be written as

$$\tan \frac{\theta_A}{2} = \sqrt{\frac{1 - a_s}{1 + a_s}} \tan \frac{\theta_B}{2}, \tag{4.216}$$

noting the trigonometric/gyrotrigonometric identity

$$\tan \frac{\theta_A}{2} = \frac{\sin \theta_A}{1 + \cos \theta_A}. \tag{4.217}$$

Replacing our gyrogeometric notation a, s by the standard notation of relativistic mechanics v, c, that is, $v = a$ is the relative speed between the Earth and the Sun, and $c = s$ is the speed of light in empty space, the stellar aberration formulas (4.215)–(4.216) are found, for instance, in [42, p. 73].

The stellar aberration formulas (4.215)–(4.216) are thus well-known in the literature, They possess an elegant symmetry in the sense that they are equivalent to the following stellar aberration formulas in which the roles of θ_B and θ_A are interchanged,

$$\begin{aligned} \sin \theta_B &= \frac{1}{\gamma_a} \frac{\sin \theta_A}{1 - a_s \cos \theta_A} \\ \cos \theta_B &= \frac{-a_s + \cos \theta_A}{1 - a_s \cos \theta_A} \end{aligned} \tag{4.218}$$

and

$$\tan \frac{\theta_B}{2} = \sqrt{\frac{1 + a_s}{1 - a_s}} \tan \frac{\theta_A}{2}. \tag{4.219}$$

The second stellar aberration formula in (4.218) is found in Einstein's 1905 paper that founded the special theory of relativity [9, p. 912]. Unfortunately, Einstein erroneously attributed the stellar aberration effect to a "Licht-quelle–Beobachter" ("Light-source–Observer") velocity, rather than to an Observer–Observer velocity as in our study, where the two observers are A (Earth) and B (Sun), as shown in Fig. 4.18. Fortunately, however, Einstein himself corrected his error in a reprint copy as noted by John Stachel in [10, p. 148, Editorial Note 10, p. 160]. The common problem in the understanding of stellar aberration within the frame of Einstein's special relativity theory is described in [38]. Fortunately, we do not face any problem in the study of relativistic stellar aberration since our study is guided by analogies with classical results in geometry and mechanics.

Interestingly, (i) the relativistic particle aberration formula (4.213) is identical with Rindler's particle aberration formula, [44, p. 86], and (ii) the relativistic photon aberration formula (4.214) is identical with the well-known relativistic electromagnetic wave aberration formula observed in *stellar aberration* [17, pp. 132–133], [41, pp. 84–87], [43, pp. 57–58], [44, pp. 81–82], [47, p. 71], [49, p. 146].

It follows from (4.218) and from the addition formulas of the cos/gyrocos and sin/gyrosin trigonometric/gyrotrigonometric functions that

$$\sin(\theta_A - \theta_B) = \sin\theta_A \cos\theta_B - \cos\theta_A \sin\theta_B = -\frac{\sqrt{\gamma_a^2 - 1} - (\gamma_a - 1)\cos\theta_A}{\gamma_a - \sqrt{\gamma_a^2 - 1}\cos\theta_A}\sin\theta_A$$

$$\cos(\theta_A - \theta_B) = \cos\theta_A \cos\theta_B - \sin\theta_A \sin\theta_B = -\frac{\sqrt{\gamma_a^2 - 1} - (\gamma_a - 1)\cos\theta_A}{\gamma_a - \sqrt{\gamma_a^2 - 1}\cos\theta_A}\cos\theta_A \qquad (4.220)$$

$$+ \frac{1}{\gamma_a - \sqrt{\gamma_a^2 - 1}\cos\theta_A}.$$

Identities (4.220) express the aberration gyroangle $\theta_A - \theta_B$ in terms of gyroangle θ_A which, for observers at rest relative to the Earth, has measure equals to that of its Euclidean counterpart, the angle θ_A, as explained in Fig. 4.1, p. 112.

Albert Stewart explains the observation of stellar aberration as follows:

"The displacement, or aberration, of starlight can be detected by direct observation [from the Earth, with no need to communicate with an observer at rest relative to the Sun] because the Earth is not always moving in the same direction. Since the Earth's orbit is nearly circular, an observer who is being carried in one direction now will be moving in the opposite direction six months later. Owing to this change in direction, starlight that is displaced one way now will be displaced the opposite way in six months from now. The actual displacement of starlight because of aberration cannot be directly observed, but the changes in this displacement can. It was through following the changes in the displacement of several stars that the English astronomer James Bradley discovered stellar aberration early in the 18th century."

Albert B. Stewart [48]

4.24   GYRO-ANALOGIES THAT MAY REVEAL THE ORIGIN OF DARK MATTER

The reader has realized that gyro-analogies between the common vector space approach to Euclidean geometry and our gyrovector space approach to hyperbolic geometry play an important role in this book and in similar explorations as, for instance, [62]. We are thus able in this book to uncover the secret world of Einstein's special theory of relativity and its underlying hyperbolic geometry by discovering gyro-analogies that it shares with the familiar Newtonian mechanics and its underlying Euclidean geometry. In this section, we present gyro-analogies that may reveal the origin of dark matter in the Universe.

4.24.1 NEWTONIAN SYSTEMS OF PARTICLES

In this subsection we set the stage for revealing analogies that a Newtonian system of N particles and an Einsteinian system of N particles share. In this subsection, accordingly, as opposed to Subsection 4.24.2, $\mathbf{v}_k \in \mathbb{R}^3, k = 0, 1, \ldots, N$, are Newtonian velocities in the Euclidean 3-space $\mathbb{R}^3$, and $m_0 > 0$ is the Newtonian resultant mass of the constituent masses $m_k > 0, k = 1, \ldots, N$ of a Newtonian particle system S.

Accordingly, let us consider the following well-known classical results, (4.222)–(4.224) below, which are involved in the determination of the Newtonian resultant mass m_0 and the classical center of momentum (CM) of a Newtonian system of particles, and to which we will seek Einsteinian gyro-analogs in Subsection 4.24.2. Thus, let

$$S = S(m_k, \mathbf{v}_k, \Sigma_0, N), \qquad \mathbf{v}_k \in \mathbb{R}^3 \qquad (4.221)$$

be an isolated Newtonian system of N noninteracting material particles the k-th particle of which has mass $m_k > 0$ and Newtonian uniform velocity $\mathbf{v}_k$ relative to an inertial frame $\Sigma_0, k = 1, \ldots, N$.

Furthermore, let $m_0 > 0$ be the resultant mass of S, considered as the mass of a virtual particle located at the center of mass of S, and let $\mathbf{v}_0$ be the Newtonian velocity relative to Σ_0 of the Newtonian CM frame of S. Then, as it is well known,

$$1 = \frac{1}{m_0} \sum_{k=1}^{N} m_k, \qquad (4.222)$$

and

$$\mathbf{v}_0 = \frac{1}{m_0} \sum_{k=1}^{N} m_k \mathbf{v}_k \qquad (4.223)$$

$$\mathbf{u} + \mathbf{v}_0 = \frac{1}{m_0} \sum_{k=1}^{N} m_k (\mathbf{u} + \mathbf{v}_k),$$

$\mathbf{u}, \mathbf{v}_k \in \mathbb{R}^3, m_k > 0, k = 0, 1, \ldots, N$. Here m_0 is the Newtonian mass of the Newtonian system S, supposed concentrated at the center of mass of S, and $\mathbf{v}_0$ is the Newtonian velocity relative to Σ_0 of the Newtonian CM frame of the Newtonian system S in (4.221).

It follows from (4.222) that m_0 in (4.222)–(4.223) is given by the Newtonian resultant mass equation

$$m_0 = \sum_{k=1}^{N} m_k. \qquad (4.224)$$

The derivation of (4.224) from (4.222) is trivial, but is presented here in order to set the stage for a gyro-analogy with its relativistic counterpart, (4.228), in subsection 4.24.2, which is far from being trivial.

The derivation of the second equation in (4.223) from the first equation in (4.223) is immediate, following (i) the distributive law of scalar-vector multiplication, and (ii) the simple relationship (4.224) between the Newtonian resultant mass m_0 of the system S and its constituent masses m_k, $k = 1, \ldots, N$.

4.24.2 EINSTEINIAN SYSTEMS OF PARTICLES

In this subsection we present the Einsteinian gyro-analogs of the Newtonian equations (4.221)–(4.224) listed in Sec. 4.24.1. The presented analogs, listed in (4.225)–(4.228) below, are obtained in [61, Ch. 11] by means of gyrogroup and gyrovector space theoretic techniques. Their proof is beyond the scope of this book and, hence, will not be reproduced here.

In this subsection, accordingly, as opposed to Subsection 4.24.1, $\mathbf{v}_k \in \mathbb{R}^3_c, k = 0, 1, \ldots, N$, are Einsteinian velocities in the c-ball $\mathbb{R}^3_c$ of the Euclidean 3-space $\mathbb{R}^3$, and $m_0 > 0$ is the Einsteinian resultant mass, yet to be determined, of the masses $m_k > 0, k = 1, \ldots, N$, of an Einsteinian particle system S.

In analogy with (4.221), let

$$S = S(m_k, \mathbf{v}_k, \Sigma_0, N), \qquad \mathbf{v}_k \in \mathbb{R}^3_c \tag{4.225}$$

be an isolated Einsteinian system of N noninteracting material particles the k-th particle of which has invariant mass $m_k > 0$ and Einsteinian uniform velocity $\mathbf{v}_k \in \mathbb{R}^3_c$ relative to an inertial frame $\Sigma_0, k = 1, \ldots, N$.

Furthermore, let $m_0 > 0$ be the resultant mass of S, considered as the mass of a virtual particle located at the center of mass of S (calculated in [61, Chap. 11]), and let $\mathbf{v}_0 \in \mathbb{R}^3_c$ be the Einsteinian velocity relative to Σ_0 of the Einsteinian center of momentum (CM) frame of the Einsteinian system S in (4.225). Then, as shown in [61, p. 484], the relativistic analogs of the Newtonian expressions in (4.222)–(4.224) are, respectively, the following Einsteinian expressions in (4.226)–(4.228) below,

$$\gamma_{\mathbf{v}_0} = \frac{1}{m_0} \sum_{k=1}^{N} m_k \gamma_{\mathbf{v}_k}$$

$$\gamma_{\mathbf{u} \oplus \mathbf{v}_0} = \frac{1}{m_0} \sum_{k=1}^{N} m_k \gamma_{\mathbf{u} \oplus \mathbf{v}_k} \tag{4.226}$$

and

$$\gamma_{\mathbf{v}_0} \mathbf{v}_0 = \frac{1}{m_0} \sum_{k=1}^{N} m_k \gamma_{\mathbf{v}_k} \mathbf{v}_k$$

$$\gamma_{\mathbf{u} \oplus \mathbf{v}_0} (\mathbf{u} \oplus \mathbf{v}_0) = \frac{1}{m_0} \sum_{k=1}^{N} m_k \gamma_{\mathbf{u} \oplus \mathbf{v}_k} (\mathbf{u} \oplus \mathbf{v}_k) \tag{4.227}$$

$\mathbf{u}, \mathbf{v}_k \in \mathbb{R}^3_c$, $m_k > 0$, $k = 0, 1, \ldots, N$. Here the relativistic resultant mass m_0, which is compatible with (4.226) and (4.227), is given uniquely by the equation

$$m_0 = \sqrt{\left(\sum_{k=1}^{N} m_k\right)^2 + 2 \sum_{\substack{j,k=1 \\ j<k}}^{N} m_j m_k (\gamma_{\ominus \mathbf{v}_j \oplus \mathbf{v}_k} - 1)}, \tag{4.228}$$

as verified in [61, Theorem 11.6].

Thus, m_0 in (4.228) is the relativistic invariant mass of the Einsteinian system S, supposed concentrated at the relativistic center of mass of S (calculated in [61, Chap. 11]), and $\mathbf{v}_0$ is the Einsteinian velocity relative to Σ_0 of the Einsteinian CM frame of the Einsteinian system S in (4.225).

4.24.3 THE RELATIVISTIC INVARIANT MASS PARADOX

In analogy with the Newtonian resultant mass m_0 in (4.224), which follows from (4.222), it follows from (4.226) that the Einsteinian resultant mass m_0 in (4.226)–(4.227) is given by the elegant Einsteinian resultant mass equation (4.228), as shown in [61, Chap. 11].

The Einsteinian resultant mass equation (4.228) presents a Paradox, called the *Relativistic Invariant Mass Paradox*, since, in general, this equation implies the inequality

$$m_0 > \sum_{k=1}^{N} m_k \tag{4.229}$$

so that, paradoxically, the invariant resultant mass of a system may exceed the sum of the invariant masses of its constituent particles.

The paradoxical invariant resultant mass equation (4.228) for m_0 is the relativistic analog of the non-paradoxical Newtonian resultant mass equation (4.224) for m_0, to which it reduces in each of the following two special cases:

(*i*) The Einsteinian resultant mass m_0 in (4.228) reduces to the Newtonian resultant mass m_0 in (4.224) in the Newtonian limit as the speed of light tends to infinity, $c \to \infty$; and

(*ii*) The Einsteinian resultant mass m_0 in (4.228) reduces to the Newtonian resultant mass m_0 in (4.224) in the special case when the system S is rigid, that is, all the internal motions in S of the constituent particles of S relative to each other vanish. In that case $\ominus \mathbf{v}_j \oplus \mathbf{v}_k = \mathbf{0}$ so that $\gamma_{\ominus \mathbf{v}_j \oplus \mathbf{v}_k} = 1$ for all $j, k = 1, \ldots, N$. This identity, in turn, generates the reduction of (4.228) into (4.224).

The second equation in (4.227) follows from the first equation in (4.227) in full analogy with the second equation in (4.223), which follows from the first equation in (4.223) by the distributivity of scalar multiplication and by the simplicity of (4.224). However, while the proof of the latter is simple and well known, the proof of the former, presented in [61, Chap. 11], is lengthy owing to the lack of a distributive law for the Einsteinian scalar multiplication (see [61, Chap. 6]) and the lack of a simple relation for m_0 like (4.224), which is replaced by (4.228).

The relativistic resultant mass (4.228) of the relativistic particle system S, (4.225), comprises of two distinct kinds of mass. These are the *Newtonian mass* m_N,

$$m_N := \sum_{k=1}^{N} m_k , \qquad (4.230)$$

which is the sum of the Newtonian masses of the constituent particles of the system S, and the *dark mass* m_D,

$$m_D := \sqrt{2 \sum_{\substack{j,k=1 \\ j<k}}^{N} m_j m_k (\gamma_{\ominus \mathbf{v}_j \oplus \mathbf{v}_k} - 1)} , \qquad (4.231)$$

which is a virtual mass originated from the internal motion of the constituent particles of the system S relative to each other.

It follows from (4.228) and (4.230)–(4.231) that the Newtonian mass and the dark mass of a relativistic system S of particles add according to the equation

$$m_0 = \sqrt{m_N^2 + m_D^2} . \qquad (4.232)$$

We have thus found that the presence of dark mass in particle systems, like a galaxy, is dictated by Einstein's special theory of relativity. Being virtual, the dark mass is dark since it does not emit light. Moreover, it does not collide and has no physical property other than mass, so that it reveals its presence only gravitationally. Hence, suggestively, dark mass could be the dark matter that cosmologists believe is necessary in order to supply the missing gravity that keeps galaxies stable [3, 6, 36, 60].

4.25   EXERCISES

(1) Use Mathematica or Maple to verify (i) (4.45), p. 126; (ii) (4.47), p. 126; (iii) (4.50), p. 127; (iv) (4.59), p. 129; (v) (4.77), p. 133; (vi) (4.79), p. 134; and (vii) (4.83), p. 135.

(2) Verify Identities (4.193) by translating [55, Theorem 8.59, p. 297] from Möbius to Einstein gyrovector spaces by means of (3.127)–(3.128), p. 89, or [55, Eqs. (6.309)–(6.310), p. 208].

(3) Carnot Theorem is well-known in Euclidean geometry. Translate it into hyperbolic geometry [8].

(4) Clifford algebras are well-known for providing a powerful mathematical language for the development of mathematical physics. Express the Möbius gyrogroup in the language Clifford algebras [12].

Bibliography

[1] Lars V. Ahlfors. *Conformal invariants: topics in geometric function theory*. McGraw-Hill Series in Higher Mathematics. McGraw-Hill Book Co., New York, 1973.

[2] Lars V. Ahlfors. *Möbius transformations in several dimensions*. University of Minnesota School of Mathematics, Minneapolis, Minn. 1981.

[3] John Bahcall, Tsvi Piran, and Steven Weinberg (eds.). *Dark Matter in the Universe*. sec. ed. Kluwer Academic Publishers Group, Dordrecht, 2004.

[4] James W. Cannon, William J. Floyd, Richard Kenyon, and Walter R. Parry. Hyperbolic geometry. In *Flavors of geometry*, pages 59–115. Cambridge Univ. Press, Cambridge, 1997.

[5] Michael A. Carchidi. Generating exotic-looking vector spaces. *College Math. J.*, 29(4):304–308, 1998. DOI: 10.2307/2687687

[6] Edmund J. Copeland, M. Sami, and Shinji Tsujikawa. Dynamics of dark energy. *Internat. J. Modern Phys. D*, 15(11):1753–1935, 2006. DOI: 10.1142/S021827180600942X

[7] Michael A. B. Deakin. The provenance of Urquhart's theorem. *Austral. Math. Soc. Gaz.*, 8(1):26–28, 1981.

[8] Oğuzhan Demirel and Emine Soytürk. The hyperbolic Carnot theorem in the poincaré disc model of hyperbolic geometry. *Novi Sad J. Math.*, 38(2):33–39, 2008.

[9] Albert Einstein. Zur elektrodynamik bewegter körper [on the electrodynamics of moving bodies] *Ann. Physik (Leipzig)*, 17:891–921, 1905. For English translation see [10]. It also appears as an Appendix in Miller [31]. Also available at URL: http://www.fourmilab.ch/etexts/einstein/specrel/www/ DOI: 10.1002/andp.19053221004

[10] Albert Einstein. *Einstein's Miraculous Years: Five Papers that Changed the Face of Physics*. Princeton, Princeton, NJ, 1998. Edited and introduced by John Stachel. Includes bibliographical references. Einstein's dissertation on the determination of molecular dimensions – Einstein on Brownian motion – Einstein on the theory of relativity – Einstein's early work on the quantum hypothesis. A new English translation of Einstein's 1905 paper on pp. 123–160.

[11] Francis C. W. Everitt, William M. Fairbank, and L. I. Schiff. *Theoretical background and present status of the Stanford relativity–gyroscope experiment*. Switzerland, 1969. In *The Significance of Space Research for Fundamental Physics*, Proceedings of the Colloquium of the European Space

Research Organization at Interlaken, Sept. 4, 1969. See also the project webcite at URL: http://einstein.stanford.edu

[12] Milton Ferreira. Factorizations of Möbius gyrogroups. *Adv. Appl. Clifford Algebr.*, 2009. in print.

[13] Richard P. Feynman, Robert B. Leighton, and Matthew Sands. *The Feynman lectures on physics.* Addison-Wesley Publishing Co., Inc., Reading, Mass.-London, 1964.

[14] Stephen D. Fisher. *Complex variables.* Dover Publications Inc. Mineola, NY, 1999. Corrected reprint of the second (1990) edition.

[15] Tuval Foguel and Abraham A. Ungar. Involutory decomposition of groups into twisted subgroups and subgroups. *J. Group Theory*, **3** (2000) 27–46. DOI: 10.1515/jgth.2000.003

[16] Tuval Foguel and Abraham A. Ungar. Gyrogroups and the decomposition of groups into twisted subgroups and subgroups. *Pac. J. Math.* **197** (2001) 1–11.

[17] Anthony Philip French. *Special relativity.* Norton, New York, 1968.

[18] Marvin Jay Greenberg. *Euclidean and non-Euclidean geometries: Development and History.* W. H. Freeman and Company, New York, 3rd ed., 1993.

[19] Robert E. Greene and Steven G. Krantz. *Function theory of one complex variable.* John Wiley & Sons Inc. New York, 1997.

[20] Mowaffaq Hajja. A very short and simple proof of "the most elementary theorem" of Euclidean geometry. *Forum Geom.*, 6:167–169 (electronic), 2006.

[21] Mowaffaq Hajja. A short trigonometric proof of the steiner-lehmus theorem. *Forum Geom.*, 8:39–42 (electronic), 2008.

[22] Melvin Hausner. *A vector space approach to geometry.* Dover Publications Inc., Mineola, NY, 1998. Reprint of the 1965 original.

[23] W. Kantor. Nonparallel convention of light. *Lett. Nuovo Cimento*, 3:747–748, 1972. DOI: 10.1007/BF02824352

[24] Hans-Joachim Kowalsky. *Lineare Algebra.* Walter de Gruyter, Berlin-New York, 1977. Achte Auflage, de Gruyter Lehrbuch.

[25] Steven G. Krantz. *Geometric function theory.* Cornerstones. Birkhäuser Boston Inc., Boston, MA, 2006. Explorations in complex analysis.

[26] J.S. Mackay. History of a theorem in elementary geometry. *Edin. Math. Soc. Proc.*, 20:18–22, 1902.

[27] G. B. Malykin. Thomas precession: correct and incorrect solutions. *Physics–Uspekhi*, 49(8):837–853, 2006.

[28] Jerrold E. Marsden. *Elementary classical analysis*. W. H. Freeman and Co., San Francisco, 1974. With the assistance of Michael Buchner, Amy Erickson, Adam Hausknecht, Dennis Heifetz, Janet Macrae and William Wilson, and with contributions by Paul Chernoff, István Fáry and Robert Gulliver.

[29] George E. Martin. *The foundations of geometry and the non-Euclidean plane*. Undergraduate Texts in Mathematics. Springer-Verlag, New York, 1982. Reprint.

[30] John McCleary. *Geometry from a differentiable viewpoint*. Cambridge University Press, Cambridge, 1994.

[31] Arthur I. Miller. *Albert Einstein's special theory of relativity*. Springer-Verlag, New York, 1998. Emergence (1905) and early interpretation (1905–11), Includes a translation by the author of Einstein's "On the electrodynamics of moving bodies", Reprint of the 1981 edition.

[32] Richard S. Millman and George D. Parker. *Geometry: A metric approach with models*. Springer-Verlag, New York, 2nd ed., 1991.

[33] David Mumford, Caroline Series, and David Wright. *Indra's pearls: The vision of Felix Klein*. Cambridge University Press, New York, 2002.

[34] Péter T. Nagy and Karl Strambach. *Loops in group theory and Lie theory*. Walter de Gruyter & Co., Berlin, 2002.

[35] Tristan Needham. *Visual complex analysis*. The Clarendon Press Oxford University Press, New York, 1997.

[36] Iain Nicolson. *Dark side of the universe: dark matter, dark energy, and the fate of the cosmos*. John Hopkins University Press, Baltimore, MD, 2007.

[37] Hala O. Pflugfelder. *Quasigroups and loops: introduction*, volume 7 of *Sigma Series in Pure Mathematics*. Heldermann Verlag, Berlin, 1990.

[38] Thomas E. Phipps. Relativity and aberration. *Amer. J. Phys.*, **57**(6) (1998) 549–551.

[39] Alfred S. Posamentier. *Advanced Euclidean geometry*. Key College Publishing, Emeryville, CA, 2002.

[40] John G. Ratcliffe. *Foundations of hyperbolic manifolds*, volume 149 of *Graduate Texts in Mathematics*. Springer-Verlag, New York, 1994.

[41] Robert Resnick. *Introduction to special relativity*. John Wiley & Sons, New York London Sydney, 1968.

[42] Wolfgang Rindler. *Essential relativity: special, general, and cosmological.* Van Nostrand Reinhold Co., New York, 1969.

[43] Wolfgang Rindler. *Essential relativity: special, general, and cosmological.* Springer Verlag, New York, 1977.

[44] Wolfgang Rindler. *Relativity: special, general, and cosmological.* Oxford University Press, Oxford, New York, 2001.

[45] K.R.S. Sastry. A Gergonne analogue of the Steiner-Lehmus theorem. *Forum Geom.*, 5:191–195 (electronic), 2005.

[46] Hans Schwerdtfeger. *Geometry of complex numbers.* Mathematical Expositions, No. 13. University of Toronto Press, Toronto, 1962.

[47] Roman U. Sexl and Helmuth K. Urbantke. *Relativity, groups, particles.* Springer Physics. Springer-Verlag, Vienna, 2001.

[48] Albert B. Stewart. The discovery of stellar aberration. *Sci. Amer.*, March:100–108, 1964.

[49] J. L. Synge. *Relativity: the special theory.* North-Holland Publishing Co., Amsterdam, 2nd ed., 1965.

[50] E. Trost and R. Breusch. Problem, and solution to problem, 4964. *Amer. Math. Monthly*, **69** (1962) 672–674. DOI: 10.2307/2310853

[51] S. M. Ulam. *Analogies between analogies*, volume 10 of *Los Alamos Series in Basic and Applied Sciences.* University of California Press, Berkeley, CA, 1990. The mathematical reports of S. M. Ulam and his Los Alamos collaborators, Edited and with a foreword by A. R. Bednarek and Françoise Ulam, With a bibliography of Ulam by Barbara Hendry.

[52] Abraham A. Ungar. Thomas rotation and the parametrization of the Lorentz transformation group. *Found. Phys. Lett.* **1** (1988) 57–89. DOI: 10.1007/BF00661317

[53] Abraham A. Ungar. Mobius transformations of the ball, Ahlfors' rotation and gyrovector spaces. In *Themistocles M. Rassias (ed.): Nonlinear analysis in geometry and topology*, pages 241–287. Hadronic Press, Palm Harbor, FL, 2000.

[54] Abraham A. Ungar. *Beyond the Einstein addition law and its gyroscopic Thomas precession: The theory of gyrogroups and gyrovector spaces*, volume 117 of *Fundamental Theories of Physics.* Kluwer Academic Publishers Group, Dordrecht, 2001.

[55] Abraham A. Ungar. *Analytic hyperbolic geometry: Mathematical foundations and applications.* World Scientific Publishing Co. Pte. Ltd., Hackensack, NJ, 2005.

[56] Abraham A. Ungar. Einstein's special relativity: Unleashing the power of its hyperbolic geometry. *Comput. Math. Appl.* **49** (2005) 187–221. DOI: 10.1016/j.camwa.2004.10.030

[57] Abraham A. Ungar. Gyrovector spaces and their differential geometry. *Nonlinear Funct. Anal. Appl.* **10** (2005) 791–834.

[58] Abraham A. Ungar. The relativistic hyperbolic parallelogram law. In *Geometry, integrability and quantization*. Softex, Sofia, (2006) 249–264.

[59] Abraham A. Ungar. Thomas precession: a kinematic effect of the algebra of Einstein's velocity addition law. Comments on 'deriving relativistic momentum and energy: II. three dimensional case'. *European J. Phys.* **27** (2006) L17–L20. DOI: 10.1088/0143-0807/27/3/L02

[60] Abraham A. Ungar. On the origin of the dark matter/energy in the universe and the Pioneer anomaly. *Prog. Phys.*, **3** (2008) 24–29.

[61] Abraham A. Ungar. *Analytic hyperbolic geometry and Albert Einstein's special theory of relativity.* World Scientific Publishing Co. Pte. Ltd., Hackensack, NJ, 2008.

[62] Abraham A. Ungar. Hyperbolic barycentric coordinates. *Aust. J. Math. Anal. Appl.*, **6** (2009), in print.

[63] J. Vermeer. A geometric interpretation of Ungar's addition and of gyration in the hyperbolic plane. *Topology Appl.* **152** (2005) 226–242. DOI: 10.1016/j.topol.2004.10.012

[64] Scott Walter. The non-Euclidean style of Minkowskian relativity. In *The symbolic universe (J. J. Gray (ed.), Milton Keynes, England)*, pages 91–127. Oxford Univ. Press, New York, 1999.

[65] Scott Walter. Book Review: *Beyond the Einstein Addition Law and its Gyroscopic Thomas Precession: The Theory of Gyrogroups and Gyrovector Spaces*, by Abraham A. Ungar. *Found. Phys.* **32** (2002) 327–330.

Index